Environmental Biodiversity

Environmental Biodiversity

By

Dr. P.R. Yadav

Lecturer
Department of Zoology
D.A.V. College
Muzaffarnagar (U.P.)

&

Dr. Shubhrata R. Mishra

Department of Botany
Vikram University
Ujjain (M.P)

D P H

DISCOVERY PUBLISHING HOUSE
NEW DELHI-110002

Reprinted - 2020

First Published - 2004

ISBN: 978-81-7141-755-1

Environmental Biodiversity

Published by:

DISCOVERY PUBLISHING HOUSE PVT. LTD.
4383/4B, Ansari Road, Darya Ganj
New Delhi-110 002 (India)
Phone: +91-11-23279245, 23253475; 43596065
E-mail: discoverybooksindia@gmail.com
discoverypublishinghouse@gmail.com
web: www.discoverypublishinggroup.com

Printed at:
Infinity Imaging Systems
Delhi

PREFACE

The present title Environmental Biodiversity deals extensively with the elements, components, natural resources, physical, chemical, geological and biological aspects of environment. The last thirty years have witnessed a vast expansion of both interest and knowledge in environmental sciences. There has been a widening public awareness of the importance of environmental interactions with particular focus on such issues as the impact of pesticides upon food chains and the loss of global biodiversity owing to habitat destruction.

The present title is intended to be accessible to any one with an interest in or responsibility for specific environmental concerns and hazards. Thus it is aimed at a broad range of readers. This book differs from the other books in its central concern with "statement about the future" as made by a wide range of scientists. The subject areas covered are extensive and specialists in every field will find materials than can be improved and extended.

This book is an effort to summarize the basic concept and principles of the subject, to present the elementary factual information with which a person to be competent in the field should be familiar, and to sow how these principles and facts may be applied in a practical way to the interests and welfare of man.

The author has freely consulted various articles, discussion notes, reviews and extracts from the various scientific journals in the preparation of the present book, in able to make it comprehensive and upto date.

Though every care has been taken by the printer, publisher and me, it is quite likely that some errors might have found their way into the book but I hope these are of very insignificant nature. However, suggestions to improve book and pointing out of errors and mistakes, if any, will gratefully be appreciated.

The author expresses grateful to her friends and colleagues whose constant inspiration have initiated her in bringing out this book.

Special thanks are given to Mr. Wasan and staff of M/s Discovery Publishing House for their whole hearted co-operation in the publication of this book.

Author

CONTENTS

Categories and Definitions, Endangerd, Vulnerable, Rare, Threatened, Out of Danger, Indeterminate, Main Causes of Threat to Species, Area Vulnerable to Rarity and Extinction of Species in India, Himalaya and Eastern India, Rajasthan and Gujarat, Andaman and Nicobar Islands, Peninsular India, Gangetic Plain, Gaps in our Knowledge Rare and Endangered Indian Plants, Action Taken to save the Rarities of Indian Flora, Future Programme of Work in the Botanical Survey of India, Endangered Fauna of India, Why Save Endangered Species, The Indian Scene, Mammals, Birds, Reptiles, Amphibians, Wildlife Conservation, Importance of Conservation, Reasons for Dwindle, Classification of Scarce Wildlife, Methods of Wildlife Conservation, History of Wildlife Conservation, Wildlife Conservation in India, Endangered Species of India, Modes of Wildlife Conservation in India, Kaziranga Wild Life Sanctuary, Mans Wild Life Sanctuary and Tiger Reserve, Man, An Endangered Species, The Future.

1

ENVIRONMENT OF FRESHWATER

The aquatic systems of the terrestrial environment are, per area, among the most productive systems of the earth. The purpose of this chapter is to examine the features and processes that contribute to their high productivity. The freshwater environments include streams and lakes, as well as wetlands such as peatlands or bogs, marshes, and swamps. Aquatic features such as lakes and wetlands depend upon climatic conditions, geological events, or biological actions to develop and maintain bodies of water. These sites are thus geographically restricted. Estuaries develop where fresh and sea water meet and are also highly localized: their location is limited mainly to the mouths of rivers Streams are far more common, ad because they are the most ubiquitous terrestrial aquatic feature on the earth, they will be the main focus of the chapter.

The Water Cycle

The water cycle, or hydrologic cycle, refers to the exchange of water between the ocean and atmosphere. On the Earth, the oceans, which cover 71 percent of the surface, are the major reservoir of the water. The oceans lose water by evaporation and gain water by precipitation. This is where the process becomes intriguing. The oceans only gain about 90 percent of the water that the evaporated. The remainder is transferred to the land and returns to the oceans as run off. The total amount of water involved in the cycle, about 0.04 percent of the Earth water, represents all of the water that appears in the atmosphere (some of which appears as rain), the rivers, lakes, wetlands, soil moisture, and groundwater. At any one time, under present climatic conditions, 97.5 percent of

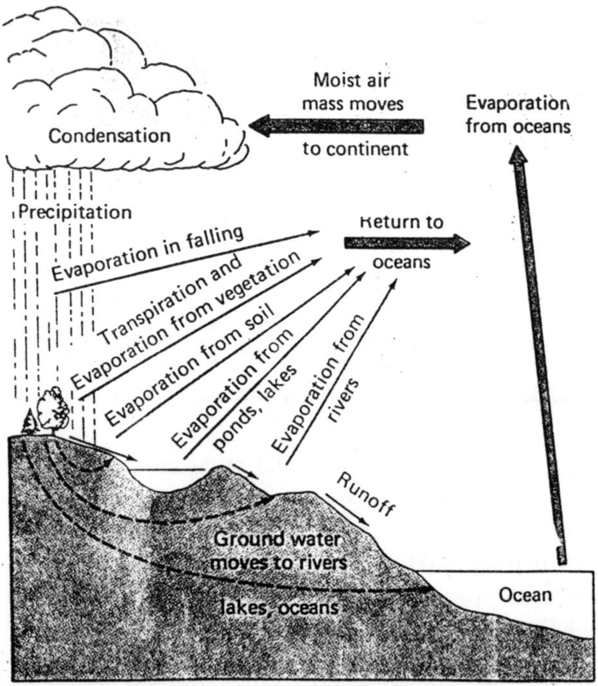

Fig. 1.1. The hydrologic cycle traces the movement of water through the atmosphere and oceans and its flow over and beneath the land surfaces.

the water on the planet occurs in the oceans. The remaining 2.5 percent is divided among the following categories. About 80 percent is in the ice caps and glaciers and about 20 percent is ground water. Lakes contain about 0.7 percent of the water soil moisture holds about 0.44 percent of the water, and the atmosphere and steams contain about 0.05 and 0.005 percent respectively of the non-ocean water. Table 1.1 lists the approximate it illustrates the time, percentage, and amount of each reservoir; residence relatively small portion of total water involved in the land part of the cycle.

The land area occupied by open water also provides perspective of the water cycle. Lakes occupy about 2.5 million km^2 1.7 percent of the surface area of the continents of the Earth (Czaya 1981). They contain about 241,000 km^2 of water. The major lakes of the Earth, *i.e.,* Lake Baikal of Asia the Great lakes of North America, and the Rift valley lakes of Africa contain almost three quarters of the freshwater. All of the remaining lakes, ponds, and wetlands share the remaining portion. Streams on the other hand occupy

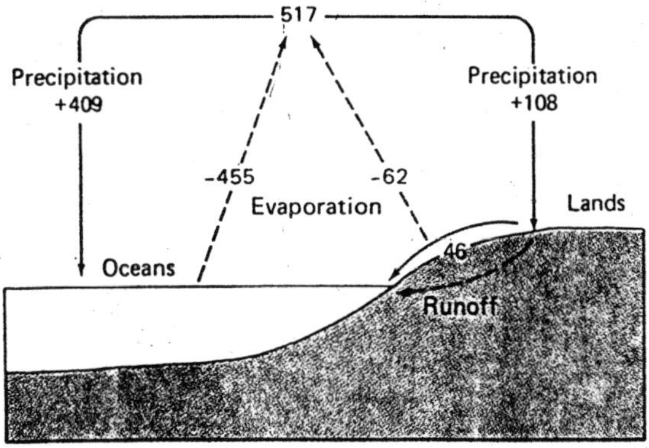

Fig. 1.2. A schematic flow diagram of the hydrologic cycle showing the estimated quantities of water that flow through the principal pathways each year. Units are thousands of cubic kilometers.

less than 0.1 percent of the area of the land (Hynes 1970). Annually, rivers, which occupy about one seventeenth of the land area of lakes, discharge about 1,000 km^2 of water back to the ocean, or nearly one tenth as much water as stored in lakes. Annually streams, with only a fraction of the depth of lakes, handle about 2.6 times more volume of water per occupied area in lakes. That in essence describes the stream–a large volume of water flowing among a narrow ribbon of band.

Table 1.1. The Major Water Reservoirs of the Earth

Reservoir	*Percent*	*Residence time*	*Amount (1,000's km^2)*
Total water	100.0		1.382,000
Oceans	97.5	1,000's years	1,347,000
Ice	2.0	1,000's–1,000,000's years	276,000
Groundwater	0.5	10–100's years	6,908
Lakes	0.0175	variable	241
Soil moisture	0.011	months	152
Rivers	0.0001	days	1.2
Atmosphere	0.001	10's of days	12.3

Streams

At any one time, streams and rivers contain about 1,230 km^2 or about 3.3 percent of the annual volume of water which returns to the seas. Stated differently, during the course of a year, the Earth's streams transport to the sea about 30 times more water than they contain at any one time. There are many possible transfers of water from streams. Some water undergoes evaporation and some steeps as ground water into the spaces between particles of sediment. Some of the ground water enters the root some to become available to plants and thence undergoes evapotranspiration.

The relationships between soils, plants, and the hydrologic cycle are described in greater detail in the next section. However, before these relationships are examined, the characteristic of streams will be considered. Streams are located at virtually all latitudes and in all climates. Climates delivers the crucial ingredients which determine the regional characteristics of streams. Streams form under conditions where precipitation exceeds evaporation and infiltration. There are several types of streams.

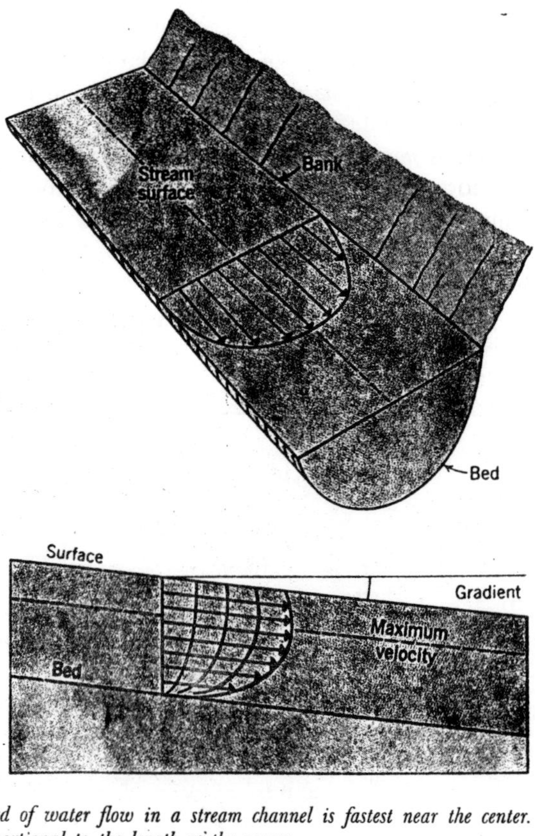

Fig. 1.3. Speed of water flow in a stream channel is fastest near the center. Velocity is proportional to the length of the arrow.

Perennial or effluent streams occur in humid climates where the supply of water is sufficient to flow in every season. The smallest drainage basins of the upper reaches of a large effluent river basins characteristically have large variations in stream flow because they have too little storage of water to continuously only a few days after each input of precipitation. Ephemeral streams occur in arid climates where infrequent precipitation and high evaporation leads to inadequate water for sustaining stream flow over long periods of time or along extended channel courses. The water in these streams tends to soak into the channel bed and such streams are called influent streams. In reality, these categories are complicated by the fact that the Earth's largest drainage basins extend across several major climatic zones.

The Drainage Basin

Stream channels occur in a seasonably well ordered pattern, most streams are part of drainage network that, connected, remove water from a drainage basin. Generally, geological conditions determine the geometry of the drainage network. This, the drainage network provides important clues to the geologic nature of the terrain. For biogeographical purposes, the most important pattern of the drainage basin shows up in the longitudinal profile of a stream. Most of the major streams and rivers of the Earth have a slope profile that is steepest at its headwaters, and has an exponential decrease in slope toward the mouth. The drainage basin of a large river consists of many smaller drainage basins containing smaller streams. A scheme has been devised to classify drainage basins according to their position within the hierarchy of the drainage

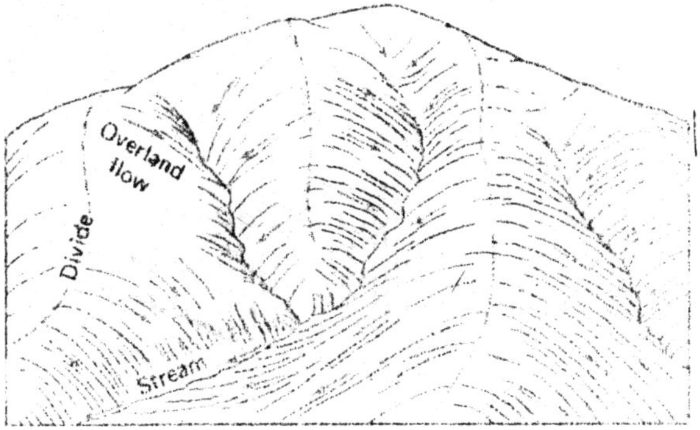

Fig. 1.4. Stream channels are fed by overland flow.

network. It is the top of the stream profile. Its slopes tend to be the steepest and with small tributary area, it tends to have low flows.

A second order stream basin is formed when two first order stream basins join; a third order stream basin occurs where two second order basin join, and so on. Higher order streams are associated with low slopes. Because of the number of streams that have joined to create high order streams, these streams usually have large volumes of water. The relationship between slope and stream flow rate provides and interesting insight into biologically significant properties of streams. Intuitively, the relationship may seem straight forward. That is, one might expect that on steep slopes, streams flow rapidly and on low slopes they flow slowly, and that the different sized streambed sediment at both segments matches expectation (*i.e.,* larger sized materials in steeper segments and finer materials in flat segments).

Streams, however, behave very differently than intuition suggests and there are many physical processes which need to be taken into account to describe the difference between upper and lower reaches of a stream. Upper reaches of a stream are found in highlands. The water source of streams is directly related to climate. Certainly

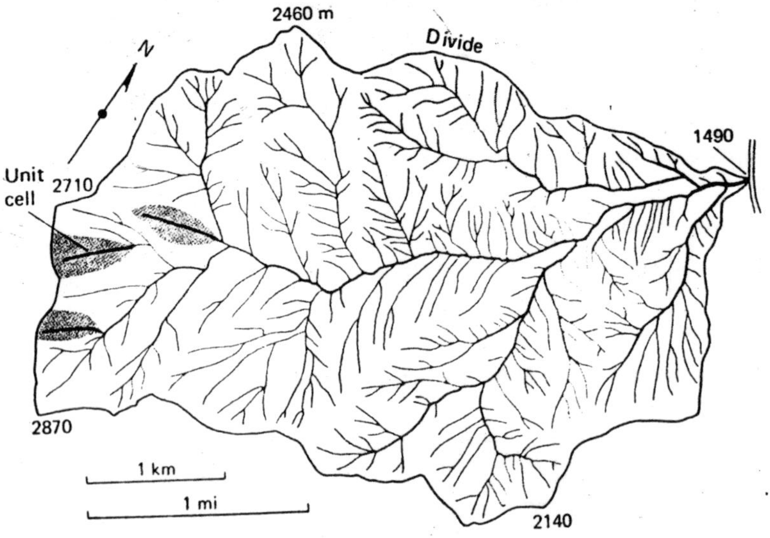

Fig. 1.5. A map of the stream network and drainage basin of Pole Canyon, Spanish Fork Peak Quadrangle, Utah. Elevations are given in meters.

orographic precipitation in upland areas is important in contributing water to streams, and depending on elevation or latitude, may be stored in the form of ice and snow and appear in the stream generally includes an assortment of boulders and other coarse materials. The valley which the stream traverses is generally a narrow, steep sided, V-shaped valley. Under these conditions, a stream encounters much resistance as it flows over the streambed. Much turbulence is apparent in the formation of various eddies and standing waves. While the water may appear to be moving swiftly, in reality, stream flow is restricted by channel resistance in it downstream progress. In the upper reaches, boulders and pebbles undergo grinding and wear as they are moved down stream (Adams 1979). The lower reaches of a stream *i.e.,* high-order stream, occur near base level.

Here, there may be expanses of flat land on either side of the stream which periodically become flooded, hence the name floodplain. The stream in these places is both wide and deep and the streambed particles consist of silts and clays. The streambed surface is thus relatively smooth, providing little resistance to streamflow which is less turbulent in appearance. In addition, the volume of the stream water in the lower reaches creates the condition whereby most of the mass of water comes into contact with only water. Only a small fraction of the volume comes into contact with the stream bottom and sides. Under these conditions, the stream velocity on low slopes is higher than it is on steep slopes in the upper reaches (Leopold et. al. 1964). Based on a continuum of bedload characteristics, slope, amount of turbulence, and flow, the stream is naturally divided into upper and lower segments. These physical divisions have distinctive properties that are important to organisms.

Regional variation of Drainage Basins

The previous section dealt with stream characteristics that generally apply to all drainage basins. There are some characteristics that distinguish drainage basins from each other. In general, most of the major distinctions occur as a result of variations in climate and geology. Climate clearly influences the characteristics of a given drainage basin. In both the Blue and White Nile headwaters, there is ample precipitation. In the headwaters, the Nile is a normal effluent stream which receives a portion of its flow from soil and groundwater. However, as the Nile flows across the wet-dry Tropics and into the desert the situation changes. The last major tributary entering the Nile is the Atbara, over 2000 km from the mouth of the Nile. From thereon, the Nile becomes an influent stream, whereby it feeds groundwater reservoirs and undergoes evaporation during its entire journey across the Sahara Desert. In this manner, runoff is continuously diminished from the flow that originates in the upper reaches of the stream.

A stream which passes through a desert for a major portion of its length is called an exotic stream. Historically, much confusion arose about the nature of streams because of the fact that the Nile appeared to come from a desert, and even more amazingly annually flooded the land even though it came from the desert. The major streams of the Earth listed in Table 1.2 are not exotic. They receive ample rainfall at their source and through out most of their course thereby gaining volume on through out to the sea. Most of the great rivers exist in humid environments.

Some of the most voluminous rivers such as the Yenisei and Lena are not located in particularly humid environments. They drain an exceptionally large regions however, and they exist in a cool climate with a relatively low rate of evapotranspiration. These drain into interior basins such as the Dead Sea, Death Valley, Caspian Sea, and Lake Eyre with the exception of Australia's Lake Eyre, these lakes are located below sea level. They all exist in climates where, overall, evaporation exceeds precipitation. These inland basin lakes of the Earth are the recipients of stream water from surrounding drainage basins. They illustrate the point that if a stream does not receive enough precipitation it cannot carve a path to the ocean.

Many interior basins of the Earth are intriguing because they contain a biota that is relict from a time when the drainage basin was connected to a larger network of streams during more humid times. These basins provide an opportunity to those presented by oceanic islands. In addition to influencing the flow volume of a stream climate plays a related role in establishing the physical conditions that cause weathering and consequently the accumulation of various minerals in soils. Tropical soils tend to be rich in aluminium and iron oxides and poor in cations such as calcium, magnesium, sodium, and potassium.

Tropical streams tend to be devoid of base in minerals and thus very soft water. During flooding the sediments carry the

Table 1.2. The Major River Basins of the Earth (Ranked by Discharge)

River	*Length (km)*	*Basin area (km^2)*	*Discharge (km^3/y)*	*Rainfall intensity (mm/y)*	*Dissolved transport (t/km^2/y)*	*Solid transport (t/km^2/y)*	*Amount transport (t × 10^6/y)*
1. Amazon	6.303	7,049,980	3,767.8	534	46.4	79.0	290.0
2. Congo	4,373	3,690,750	1,255.9	340	11.7	13.2	47.0
3. Yangtze	5,529	1,959,375	690.8	353	NA	490.0	NA
4. Mississippi-Missouri	6,215	3,221,183	556.2	173	40.0	94.0	131.0
5. Yenisei	4,991	2,597,700	550.8	212	28.0	5.1	73.0
6. Mekong	4,186	810,670	583.3	664	75.0	435.0	59.0
7. Orinoco	2,066	906,500	538.2	594	52.0	91.0	50.0
8. Parana	3,874	3,102,820	493.3	159	20.0	40.0	56.0
9. Lena	5,853	2,424,017	475.5	196	36.0	6.3	85.0
10. Brahmaputra	1,610	934.990	475.5	509	130.0	1,370.0	75.0
11. Irrawaddy	2,152	431.000	443.3	1,029	NA	700.0	NA
12. Ganges	1,610	488.992	439.6	899	78.0	537.0	76.0
13. McKenzie	3,276	1,766,380	403.7	229	39.0	65.0	NA
14. Ob	5,508	3,706,290	395.5	107	20.0	6.3	50.0

15. Amur	4,347	1,843,044	349.9	190	10.9	13.6	20.0
16. St. Lawrence	2,512	1,010,100	322.9	320	51.0	5.0	54.0
17. Indus	2,898	963,480	269.1	279	65.0	500.0	68.0
18. Zambezi	2,737	1,329,965	269.1	202	11.5	75.0	15.4
19. Volga	3,689	1,379,952	256.6	186	57.0	19.0	77.0
20. Niger	4,186	1,502,200	224.3	149	9.0	60.0	10.0
21. Columbia	1,955	668,200	210.8	316	52.0	43.0	34.0
22. Danube	2,861	816,990	197.4	242	75.0	84.0	60.0
23. Yukon	3,186	865,060	193.8	224	44.0	103.0	34.8
24. Fraser	1,368	219,632	112.4	512	NA	NA	NA
25. San Francisco	3,199	652,680	107.7	165	NA	NA	NA
26. Hwang-Ho (Yellow)	4,671	1,258,740	104.1	83	NA	2,150.0	NA
27. Nile	6,693	2,849,000	80.7	28	5.8	37.0	10.0
28 Nelson	2,576	1,072,260	76.2	71	27.0	NA	31.0
29. Murray-Darling	5,427	1,072,908	12.6	12	8.2	30.0	2.3

Note : NA=not available.

aluminium and iron oxides into tropical rivers. Like wise, middle latitude dry soils are enriched in a variety of base elements. The streams of these regions tend to be rich in base elements and thus are hard water. During flooding, the stream sediments are enriched in base rich elements. In humid middle latitude climates, the solid may be heavily leached and the waters arising from them poor in base elements and therefore soft. High latitude soils are rich in silica and the streams reflect this fact. Streams may tend to mirror soil-forming processes but generalizations do not reflect the complexity of the actual chemical characteristic of streams.

Streams are strongly influenced by regional or local geologic conditions. The water quality of the stream is determined by minerals with which it comes into contact. For example, the Colorado River passes through a salt dome in its course to the Gulf of California. In doing so, the river picks up sodium chloride. In an arid climate, the sodium chloride becomes concentrated during evaporation, and this influences water quality and the type of organisms that can exist in the river, especially during droughty years. Some rivers flow through limestone formations and pick up calcium carbonate. The dissolved carbon dioxide (CO_2) in water readily reacts with calcium carbonate to form soluble calcium bicarbonate. Calcium bicarbonate, however, stays in solutions only as long as there is a certain level of free CO_2. Any condition, such as photosynthesis, which removes carbon dioxide from water also causes the precipitation of calcium carbonate.

The calcium carbonate precipitate builds up a crystalline deposit around cells of algae. These deposits of calcium carbonate are called travertine (Hynes 1970). Such rivers have an unusual flora and fauna. Apart from physical conditions that arise from various regional and geological influences, drainage basins also have characteristics that result from organic activity within the basins. The next sections will further develop the connections between organisms and drainage basins.

Soil Plant Atmosphere Continuum

The soil plant atmosphere continuum refers to the many interconnections between the physical and organic components of the landscape. The atmosphere provide the major energy and hydrologic input and also creates the conditions (wind, water vapour, and heat) for a large percentage of the output (evaporation) from a location. Plants respond to the conditions provided by the atmosphere, although the atmosphere too is influenced by albedo (the reflection of light energy from the Earth's surface) and water vapour handling properties of plants.

Soils play a significant role in the hydrological cycle because of their water-holding capacities. Due to their requirement for water, various plants are both influenced by water availability and influence the amount of water in soils. Plants also play a role in soil formation. The drainage basin has been the key unit of the landscape that has provided opportunities for experimental observations of the relationships between soils, plants, and the atmosphere, and their influence on the movement of water.

The Soil Factor

The basic point of reference to beg... .. discussion of the movement of water in the landscape is soil. Soil has varying capacity to absorb water. Infiltration of water depends upon several physical properties of soil, Particle size is the most important. If the particles are sand sized, water generally infiltrates rapidly. Smaller particles such as silt and clay tend to pack together and have much reduced infiltration rates. The existence of roots, soil fauna, and organic matter can moderate the infiltration rate considerably. Because of the relationship between volume and surface area, particle size strongly influences water-holding capacity. Larger particles such as sand adsorb water freely and then, because of a lack of surface area, be unable to retain soil moisture.

Clays generally have a higher water-holding capacity because of the large surface area of the clay particles. But, if infiltration is poor, and spaces between clay particles too confined, then water-holding capacity is also reduced. Silts and loams (that is, mixtures of clay and sand-sized particles) have the largest amount of available water holding capacity reserve for plant growth. Three different types of bonding between water and soil particles are important to plants.

Gravity water is that which occurs in liquid state and fills the spaces as it permeates the soil. This water is freely available to plants and moves through the soil under the influence of gravity. Capillary water clings to small particles and fills small spaces of the soil. It is also available to plants as it slowly moves into the water table. The third type of water is called hygroscopic water. It is so tightly bound to soil particles by molecular forces that it is unavailable to plants which wilt without other water. This water does not move through the soil (Brady 1974). Soils releases water to streams and groundwater when two conditions are met.

The soil must have sufficient water, and the water must be under the influence of gravity. When precipitation occurs at a rate that is equal to or greater than the rate of soil infiltration, the excess water flows under the influence of gravity as runoff toward the stream. If more water infiltrates the soils than can be held by it, the surplus flows as ground water and reappears in springs and streams, or enters an underground aquifer.

The Vegetation Factor

No discussion of the water cycle or the properties of a drainage basin is complete without inclusion of the organic processes and in particular plant modification of the movements of water. Plants influence the operation of the drainage basin to such a large extent as to be inseparable from it. The effect of vegetation on the quality and yield of water from a drainage basin is a topic that has received considerable attention (Colman 1953). One recent study is the Hubbard Brook Experimental forest study (Bromann & Likens 1979). The experimental design of the research made it possible to determine the degree of management of drainage basin commodities by the biotic community. The commodities in this case include water and the major plant nutrient both in the soils and the atmosphere.

A major purpose of the research was to measure the exchanges between bedrock, soil, plans, the atmosphere, and rivers. The research was conducted on six drainage basins in the White Mountain National Forest in New Hampshire. The drainage basins rest on impervious gneiss bedrock. The studies began with the measurements of the water budge of the six undisturbed drainage basins, from 1955 through 1968. Though the use of specialized dams designed for measuring the flow rare of streams, runoff was measured. Also, the settling ponds behind the dams were used to measure sediment movement. Provisions were made to measure precipitation, as well dissolved mineral load of both incoming precipitation and outgoing runoff waters. The major water budget results of Hubbard Brook Forests are as follows:

Measured inputs and outputs

Average annual precipitation	= 123 cm
Average annual runoff	= 72 cm
Difference	
Average annual evapotranspiration	= 51 cm

From these results it can be concluded that the combination of evaporation (i.e., water loss from all wetted surfaces) and transpiration (water movements from soil and groundwater to the atmosphere via plants) accounts for 41 percent of the total annual amount of water entering the water basin. The removal of water from the soil by plants is relatively straightforward. The amount of energy required to extract soil water is small compared to the amount of energy needed to evaporate water from the leaf. Thus, up to the wilting stage, all soil water can be assumed to be equally available. Provided there is soil moisture available to a plant, hydrostatic pressure within plant conducting tissues reaches maximum values.

Stomata are openings in the leaf that control leaf ventilation. When light is not limiting the stomatal apertures of the leaf open to the maximum. In this way the leaf obtains the greatest possible flow of carbon dioxide. However, this condition also allows the rapid outflow of water vapour. Because of the limitations in the amount of carbon dioxide which a plant can obtain from the atmosphere, plants use much more water during ventilation than they actually require for physiological purposes (Penman 1970). In the humid mid-latitude climate of New Hampshire precipitation is relatively evenly distributed throughout the year; runoff on the other band, is unevenly distributed. Most of the runoff occurs during March through May. Peak runoff is attributed to the spring melt of accumulated winter snow. Thus, 57 percent of the total runoff occurs during one quarter of the year. April alone accounts for 35 percent of the total when rain often falls on top of the snow. At the other extreme, during August only 0.7 percent of the annual average runoff takes place.

The extremely low runoff during the mid to late summer can be explained by the effect of plant growth and transpiration on soil and ground water (Borman and Likens 1970). In spring as the air and soil temperatures warm up, the vegetation begins to leaf out, ending the period of winter dormancy. Evaporative losses of water increases with rising temperatures and plants begin to draw soil and groundwater at rates which exceed replenishment by precipitation. During the peak temperatures of summer, soil and groundwater are being withdrawn by plants to the extent that almost no water is available for runoff. During low flow, small streams undergo a diurnal oscillation.

During daylight, streamflow drops as evapotranspiration increase and during darkness streamflow increases. In autumn, as temperature decreases, large reductions occur in plant transpiration as plants shed their leaves and become dormant again. At this time of the year, rainfall replenishes the soil and groundwater reserves and runoff from the water increases. The ground freezes and snowfall accumulates during the winter leading to repeat of the sequence of the annual runoff cycle. In the Hubbard Brook study, the role of plants in modifying the movements of water was further demonstrated by deforestation of one of the six drainage basins. Following deforestation, the annual runoff of the drainage basin increased by about 30-40 percent over the amount of runoff of the previously forested drainage basin.

The greatest differences occurred during the spring and summer months when runoff values were about 400 percent higher than expected from the basin in its vegetated state. The difference represents the amount of water which was normally transpired by vegetation in an undisturbed drainage basin. Unvegetated drainage -basin runoff becomes more evenly distributed during the year because it closely reflects precipitation and becomes disconnected from the influence of plant processes (Likens et. al 1970.) Another major thrust of the Hubbard Brook study was on the biogeochemical cycle of a forested drainage basin. Comparison of disturbed and undisturbed basins demonstrated the degree of control that organisms exert on the landscape. The chemical quality of both precipitation (input) and runoff (output) were studied on the forested and disturbed drainage basins. Table 1.3 lists the precipitation inputs and runoff elements in outputs of elements in undisturbed and disturbed basins.

Table 1.3. Comparison of the Chemical Inputs and Outputs (kg/ha) of Undisturbed and Disturbed Drainage Basins

	Forested		*Disturbed*	
Elements	*Precipitation*	*Runoff*	*Precipitation*	*Runoff*
Calcium	2.6	11.8	2.3	77.3
Sodium	1.5	6.9	1.3	18.1
Magnesium	0.7	2.9	0.6	16.2
Potassium	1.1	1.7	0.5	23.0
Nitrates	4.6	1.3	5.8	104.0

In the forested drainage basin, the ratios between the input and output of the elements calcium sodium magnesium and potassium range from 0.2 to 0.6. In the case of nitrates the atmospheric contribution is about 3.5 times larger than the amount appearing as runoff. An important discovery of the research was that the loss of cations i.e., the nutrient ions of Table 1.3 is balanced by the input of positively charged hydrogen ions delivered by precipitation. A large proportion of the hydrogen ions are associated with sulphates and nitrates that apparently originates with industrial activities. When the water containing hydrogen ions enters the soil, the hydrogen ions replace the nutrient cations attached to the exchange sites of the soil, and the nutrients are flushed out of the soil with runoff. Water quantity and quality variation due to the disturbance can be more fully appreciated by comparing drainage-basin outputs before and after deforestation as shown in Table 1.3. Several key points can be made.

Most striking is the fact that the disturbed drainage-basin biota loses control over the movement of elements in a local region. All of the nutrient elements were being released at a rate which was between 14 and 46 times greater than gained via precipitation. Many other changes were also observed in the stream system. Physical qualities of the water undergo fundamental changes. In the forested basin, the steams are in deep shade during summer. As a result there is a very small daily fluctuation of temperature of the seam water. Within a disturbed drainage basin direct sunlight warms the soil during daytime and radiation losses during night cool the runoff. The overall change in warmer runoff which undergoes relatively large diurnal temperature fluctuation. It was also observed that the release of nutrients into the stream in large quantities (or eutrophication) caused severe deterioration of stream water quality.

The combination of large quantities of nutrients and warm water triggered the rapid growth of algae in the stream. The algae in turn died and sank to the bottom and the bacterial decay processes consumed large quantities of oxygen. The stream became anoxic and unable to support higher forms of aquatic life. In conjunction with eutrophication there was increased erosion. In the deforested drainage basin, without a source of leaves, the layer of litter was lost from the land surface and stream bottom. These leaf-litter substances are extremely important to the stream because they from a biological barrier to erosion. Leaf liter loss, coupled with increase runoff, caused a significant increase in stream channel erosion.

In sum, the drainage basin functions as a whole system in which plant and their litter play a critical role in tying the parts of the system together (Likens et.al. 1970). Experimental work in several locations has revealed that not only is the drainage basin influenced by plants but that the amount of streamflow is a function of the species of plants present in the basin. One long term study has been conducted in the southern Appalachaina mountains of southwestern West Virginia (Swank and Douglass 1974). The naturally occurring deciduous oak-hickory forest was cut and replaced with a stand of white pine (Pinus strobus). The study demonstrated the physical differences between basins vegetated with different species of plants. The soils within a basin vegetated with conifers tended to be much drier than those vegetated with hardwoods. Conifers photosynthesis and evaportanspire much earlier in the spring than do broadleaf deciduous trees.

One measure of leaf surface is the leaf area index, which is a measure of the area of leaf surface compared with the area of ground the plant is occupying. During the dormant season conifers have a much larger leaf area index than broadleaf deciduous trees (9.9 compared to less than 1) and thus intercept and subsequently evaporate much more rainfall. Overall, the evaporation and transpiration by conifers results in a stream-flow reduction of 20 percent compared to drainage basins composed of the broadleaf deciduous forest (Swank and Douglass 1974). The reduced soil moisture resulting from higher transpiration rates among conifers also limits the growth of ground cover. The reduction in ground cover resulted in changes which influenced flooding cycles and water quality.

During heavy precipitation, runoff of the sparsely vegetated surfaces caused an increase in soil erosion. Stream channels were depended and slopes became steeper as an adjustment to the influence of a conifer forest. Also the natural acidity of pine needle resulted in the leaching of cations and hence water contained greater quantities of various element. Type of vegetation controls basic characteristics of soil moisture, ground cover, and chemical regime, which in turn influence drainage basin characteristics such as water quality, sedimentation transport, channel depth and angle of slope. Vegetation cover can influence local water availability by acting as a cloud droplet impaction network which deposits water and aerosols. Examples of this phenomenon are reported from several elevations of the northern Appalachians (Lovertt et. al. 1982) and Lomas de Lachay in coastal Peru (Holdridge 1967). In the subalpine forests of the northern Appalachians, input from cloud droplets account for about 46 percent of the water, and from 150-430 percent more chemical inputs than bulk precipitation. In the Peruvian forest, precipitation recorded under the tree-crown yielded 17 percent more water than was recorded in the open.

The mechanism that causes this increase in precipitation is the condensation of moisture or deposition of cloud droplets on the aerial portion of the tree. The leaves act as large nets that capture droplets of moisture which do not otherwise have a possibility of coalescence to form precipitation. Thus process attests to the capacity of the tree canopy to influence and enhance its own microclimate. When all aspects of the relationship between plants, albedo, precipitation, and soil water are taken into consideration it becomes apparent that vegetation plays a major role in modifying the Earth's climate.

Soil Moisture Balance

The above discussion of the relationship between precipitation, vegetative cover, soil moisture, evaporation, transpiration, and runoff provides reason to consider alternative methods of classifying climate beyond the simple use of temperature and precipitation. C.W. Thornthwaite (1948) developed a method for classifying climates on the basis of energy and moisture criteria. Thornthwaite's soil water balance analysis uses month-by-month values of precipitation, potential evaporation, actual evaporation, and soil moisture surpluses or deficiencies to determine the water available for plant transpiration and growth at a given locality. Potential evaporation (PE) is the amount of soil moisture which could be used by the vegetation if there was no limit to the supply of water. The principal determinant of PE is solar radiation.

PE is usually expressed as the depth of liquid water that could be transpired per month. The value changes with seasons. When soil moisture is depend and only hygroscopic water remains, the estimated actual rate of evaporation (AE) will fall below the potential rate of evaporation. By using these two estimated moisture parameters, Thornthwaite was able to chart the local deficit or surplus of soil moisture. Several problems have been observed in the calculation of water balances in various places. Calibrations have been developed to enhance the reliability of the Thornthwaite

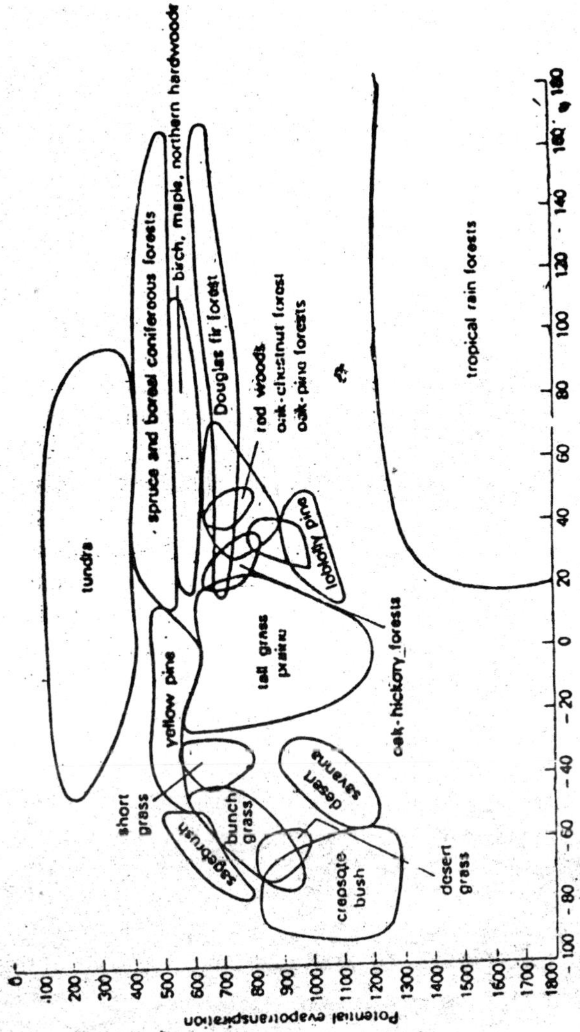

Fig. 1.6. Natural communities as a function of potential evapotranspiration and moisture index.

method in localities where they need adjustment. One particularly useful measure for evaluating the long-term availability of water resources is the moisture index.

The moisture index is calculated in two ways. It is the ratio of the difference between water surplus and deficit and the potential evaporation 100 [S–D)/PE]; and it is the ratio of the difference between precipitation and potential evaporation and the potential evapotranspiration 100 [P/PE)–1]. In general, when the potential evaporation of stations bearing particular natural vegetation-types is plotted against the moisture index of the same stations, different vegetation associations become distinguishable from one another. A forest requires in index of about 20, while grasslands and desert exist where the moisture index is negative. Likewise, the great rivers of the Earth arise from regions with positive moisture indices, while inland basins are associated with negative moisture indices.

General Characteristic of Stream Organisms

Comparisons have been made between stream organisms and their marine counterparts (Patrick 1975). In general, stream organisms as a group are smaller than their marine relatives. For example, porpoises of the Amazon though fairly large for stream organisms, are smaller than marine porpoises. Also, it has been observed that fish living in large rivers are usually larger than fish inhabiting small streams. The small size of stream organisms is probably related to the conditions found in the small streams. Narrow channels and seasonal fluctuations promote the survival of small as opposed to large animals. Small streams undergo greater fluctuations in water due to small catchment area and lack of storage capacity. Large numbers of small animals have a greater likelihood of surviving the inevitable fluctuation of small streams. Under these unstable conditions small numbers of large freshwater animals would be comparatively prone to extinction. In one way, freshwater environments of the Earth are remarkably similar. They support the same four types of insects: water striders, whirligig beetles, backswimmers, and springtails.

Water striders glide or dart across the water surface by rowing with their middle pair of legs. Gerris is a genus found on freshwater surfaces in all parts of the Earth. Whirligig beetles (Gyrinues and Dineules) live half in the water and half out, set up ripples as they move, and can dive or fly, and crawl on solid surfaces. Backswimmers (Nolonecta) live in ponds and the relatively quiet pools of streams. They swim belly-up and live on the underside of the surface film of the water. Their major food is the midge or mosquito. The fourth cosmopolitan surface-film aquatic insect is the springtail (Padura and Sminthurus). These latter are the most ancient surface-water insects and first appeared about 375 Ma ago. They alone inhabited the surface film until 250 Ma ago. Springtails feed on plant spores that settle onto the water surface. Stream organisms generally have short life cycles compared with deep lake or marine organisms.

Many species have life cycles that last only one day. Since life cycles are short, the reproductive rate tends to be very high. Stream communities tend to have long food chains that include four or five transfers of energy. Algae are important both as a food source and as a substrate for the growth of other epiphytic food sources. Rooted plants are not as important as a food source as they are as a substrate or shelter for other organisms. They thus serve to increase habitat diversity (Patrick 1975). Streams commonly contain very numbers of species.

Many species have developed specialized organs or appendages for surviving in currents (Vogel 1981). Hard water steams tend to support fewer species of algae, larger number of protozoa and insects and slightly fewer species of fish than soft-water streams (Patrick 1975). Along a single river, the number of specie in common during a single collection time between different collection sites tends to be higher than the number of species in common found at a single site during different collection times (Patrick 1975). Overall there is a high species turnover in time and space. Examination of the physical characteristics that influence biotic qualities of streams provides insight into the biogeographic patterns of the stream habitat.

Photosynthesis and Respiration Ratio of Stream

The traditional way of considering a community of organisms is to regard them as a locally self-sustaining entity to which radiant energy is supplied. The energy passes through the various trophic levels, that is, the primary producer, primary consumer, secondary consumer levels, and so on, and the decay chain breaks down the organic materials so that the constituent elements can be subsequently returned to the organisms to be recycled. This biogeochemical model may be a reasonable conceptual framework for many terrestrial and even lake communities, but it is inappropriate for the stream system.

In the stream, everything released into solution moves down stream and cannot be locally recycled (Hynes 1970). A generalized approach that succeeds in conveying the dynamic features of the stream is based on the longitudinal profile of the stream. There are physical features of the stream which change from the headwaters to the mouth that influence the ratio of photosynthesis and respiration. Some stream segments have almost no photosynthesis and thus a ratio of photosynthesis and respiration (P/R ratio) less than 1. The organic matter, usually leaves or needles, which supports the growth of organisms here must be imported into the stream segment. These segment are principally the domain of heterotrophic organisms. Other segments of the stream may have more photosynthesis than respiration and thus a P/R ratio greater than 1, and large numbers of autotrophs (Cummins 1975). The physical and biological properties of the upper and lower segments of a large stream ae sufficiently different that the two parts of the stream have been subdivided into a to rhythron and potamon.

Rhythron

The turbulent flowing stream in the elevated area i.e., head waters and mountain streams, of drainage basin is known as the rhythron (also called rhythron), literally flow plus elevation. The general changes occurring in the longitudinal profile of a stream and their influence on stream organisms. The rhythron is the coldest part of the stream from the source to the point where the monthly mean temperature rises to about 68-75°F (20-24°C), where water flow is turbulent, oxygen concentration is always high, and the streambed is composed of rocks, stones, or gravel. The head-water region of a stream is typically densely vegetated and, because of the narrowness of the stream, dimly lit.

The food chain of the headwater consists of inputs of organic matter from the vegetation via leaf fall. Because the stream valley is typically steep sided in the headwaters, the quantity of leaves per cubic meter of stream per year far exceeds the quantities in lower reaches of the stream. The nutrient values vary with the type of vegetation. Species of Alnus (alder) have root nodules and symbiotic relationship with bacteria similar to legumes. Thus, Alnus leaves contain about four times more nitrogen than other deciduous species. Confiners are relatively poor nutrient supplier to stream (Hynes 1970). The rhythron supports large numbers of benthic (bottom dwelling) organisms specialized shredding organic matter and another group specialized in capturing course particulate organic matter (CPOM). The CPOM in turn is further digested by these organisms and becomes line particulate organic matter (FPOM) of the water and sediments. The shredders obtain their food source from microorganisms and hyphomycete fungi growing on leaf tissue (Williams 1981).

One group of shredders that are important are aquatic hyphomycete fungi of the genus Ingoldia (=Gyoerfyella) Which colonize leaves that fall into stéam during the autumn (Barlocher 1982). Along with shredder ae collectors which either trap the FPOM as it passes in the water or collect it from sediments. Caddisflies spin net which capture suspended organic matter. The most ancient caddisfly taxa live in turbulent currents (Alstead 1982). Other organism feed on the microorganisms that colonize the surfaces of organic particles and thus return the particles, minus microorganisms back to the FPOM pool to be utilized repeatedly by other benthic organisms down stream. Because of the fact that primary productivity is low grazers are poorly represented. Predators such as stoneflies and fish species such as salmonids and sculpins are also present in these well aerated, organically rick, high protein rhythron waters.

Potamon

As one descend along the longitudinal profile of the steam, several physical and biological changes occur. The potamon (means river) is the lower stream segment where the monthly mean temperature rises to over 68-75°F (20-24°C), where oxygen deficits occur at times, where flow quickens and the channel deepens, and where the bed consist mainly of sand, silt or clay. There is a gathering of water resulting from the confluence of additional streams, and the stream broadens to the extent that the middle becomes exposed to sunlight. In humid climate this happens when streams become-third-or-fourth-order.

The character of the stream shifts from a predominantly heterotrophic to an autotrophic condition. Planktonic forms are common in the surface waters, and in the benthic community, grazers and collectors predominate. Shredders become less common. The caddisflies of more modern taxa have evolved to live in the laminar currents which have greater concentrations of particulate food (Alstead 1982). Predators include both invertebrates and vertebrates. Collector species dominate the river benthos as a result of the inflow of large amounts of FPOM produced up stream, and locally produced FPOM from planktonic forms. Overall, the P/R ratio becomes less than 1 because high turbidity blocks sunlight. However, high turbidity may be a modern problem because many early accounts of streams along the east coast of North America indicated clarity where now there is turbidity (Trimble 1974). The transition zone between the cooler, turbulent, shallow, and aerated rhythron and the warmer, deeper, smooth-flowing potamon generally has the largest number of species because of overlap in the ranges from both segments of the stream.

The total number of fish tends to decrease while the weight per unit fish tends to increase in the downstream reaches of the stream. The weight per unit area remains more or less constant in the stream (Hynes 1970). North American rivers have many more species of fish than do northwest European rivers, through there are many similarities between the two funas. Glaciation had an unequal influence on the faunas of the two areas. In North America, as the climate change, the river fauna was able to retreat to warmer climates in the south. In north western Europe, species were caught between the ice sheet building in the north and the glaciers forming in the Alps, resulting in the extermination of many species (Hynes 1970). The above discussion is based on research conducted primarily in humid mid-latitude climates.

There are many factors which influence the productivity of a stream. For example the seasonality of precipitation and streamflow changes with such factors as altitude, latitude, continental position, that is, all of the variables that influence climate. These same factors, because of their role in productivity, also influence the amount of organic matter available in a stream. In addition, the composition of vegetation, whether, coniferous or broadleaf deciduous forests (and even the species such as alder or oak), also affects the amount of organic matter in a stream. This brings the discussion back to the water balance idea discussed earlier in the chapter.

A comprehensive view of the complexity of the stream biosystem requires combining the principal factors which govern the operation of the regional water balance with factors that influence litter production and organic matter availability in the stream. The average timing of streamflow, waterfall, and peak particulate organic matter transport into streams is shown for several North American regions. The influence of latitude is equally profound. Middle-latitude steams include rhythronic and potamonic zones and support both cold and warm water adapted species. Potamonic species are capable of surviving–but not thriving–in rhythronic waters. Rhythronic animals cannot survive in the warm, oxygen-depleted waters of the potamon (Williams 1981).

In high latitude environments, the streams are typically rhythronic, supporting species adapted to cold, well-aerated water throughout the entire length. In tundra freshwater systems, organic matter has accumulated in the form of terrestrial peat, and is transferred to higher organisms via aquatic insect larvae (Schell 1983). In lower latitude streams, the rhythron may be very short or absent, with the potamon dominating the river. Large tropical lowland rivers have distinctly high fish species diversity, and the fish are characterized by having many adaptations to low oxygen supply (Welcomme 1979).

The Amazon undergoes seasonal fluctuations of as much as 20 m (Gentry and Lopez-Parodi 1980) and floods large areas during high water stages. The total efflux of organic carbon for the Amazon has provided a basis for estimating the total organic carbon input to the oceans from the Earth's river. The Amazon contributes about 10^{14} g of carbon and the total for the Earth's rivers must exceed 10^{15} g (Richey et. al. 1980). One source of organic carbon is from swamp hardwood forests and floodplains. This organic carbon enters the river during flooding. The influx of organic carbon tends to be moderated by biological processes because the carbon promotes increased organic activity which in turn removes organic carbon from the river at a faster rate. The overall results is a relatively constant transport of organic carbon (Richey et. al. 1980).

The flood waters of the Amazon open up extensive habitats to aquatic organisms. During high water periods the forested lowland flood plains are inundated, forming extensive varzea forests. These forests are the base for the aquatic food chain. Many common fishes move into the inundated forests and voraciously consume fruits and seeds. During low water period, the same fish move back into the main river channel or floodplain lakes and consume little food (Goulding 1981). There are many fascinating examples that provide insight into the biogeographic patterns of streams and organisms.

Natural Impoundments

Several physical processes form water holding depressions in the landscape. These include glacial action stream meandering tectonic forces and structural weaknesses. During Pleistocene times, continental glaciers formed lake-pitted kame and kettle landscapes in many areas of the Northern Hemisphere, such as the lake areas surrounding the Great Lakes of North America or dotting the landscapes European countries around the Baltic sea. Enormous blocks of ice were buried in glacial till and when they melted left depressions or kettles which filled with water, as well as ridges which are known as kames. Rivers play a role by creating oxbow lakes as they cut and fill their sedimentary banks and undergo course changes in their flood-plains.

Tectonic processes have formed the intermountain basins in Western Northern America as well as coastal subsidence in the Atlantic and Gulf coasts of the continents. Structural weakness that have resulted from the weathering and dissolution of the underlying limestone formations have created sinkholes depressions in several regions such as south Florida.

Rift Lakes

The rift valley lakes of East Africa are spectacular for their size and for the diversity of their fish fauna (Fryer & Iles 1972, Beadle 1981). Lake Malawi is the southern-most rift valley lake; it is 620 km long, 80 km wide and has a single large basin which is 750 m deep. Lake Tanganyika is even larger and deeper and has three basins but has a smaller diversity of fish. Lake Malawi has more, species of fish than any other lake in the world. About 300 species of fish are currently known by many new species are being described, and it is estimated that at least 1,000 species probably exist in the lake (Ribbink *et.al.,* 1983). The most fascinating biological feature about Lake Malawi is that nearly 95 percent of the fish are members of a single family, Cichlidae.

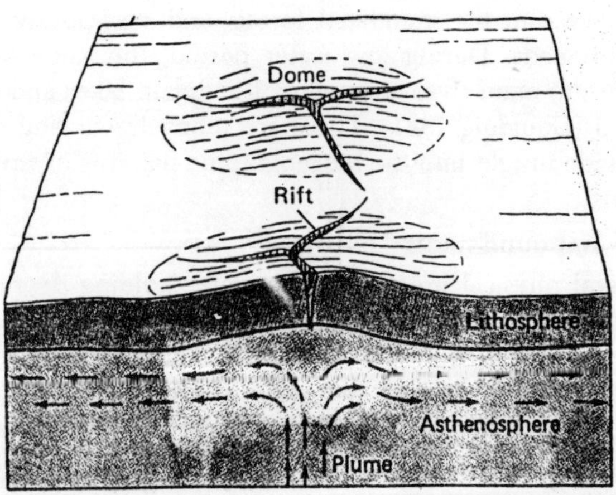

Fig. 1.7. Domes formed by lifting of the lithosphere over mantle plumes are fractured into three-armed tension rifts.

Cichlids are a large family of freshwater fish and are found in Southern Africa, Srilanka, India, South America, and one species is found in Texas. Their distribution (with the exception of the member found in Texas) suggests a long history which extends back to a time when the contents of Gondwana were connected. All but four of the cichlids found in Lake Malawi are endemic to the lake. All species except one are mouth brooders. The exception is the red bellied bream, Tilapia rendalli, which digs and broods in tunnels. The density of cichlid species is truly amazing. On Domwe and Thumba Island near Cape Maclear, in Lake Malawi National Park, there are rocky areas where densities of 15 to 20

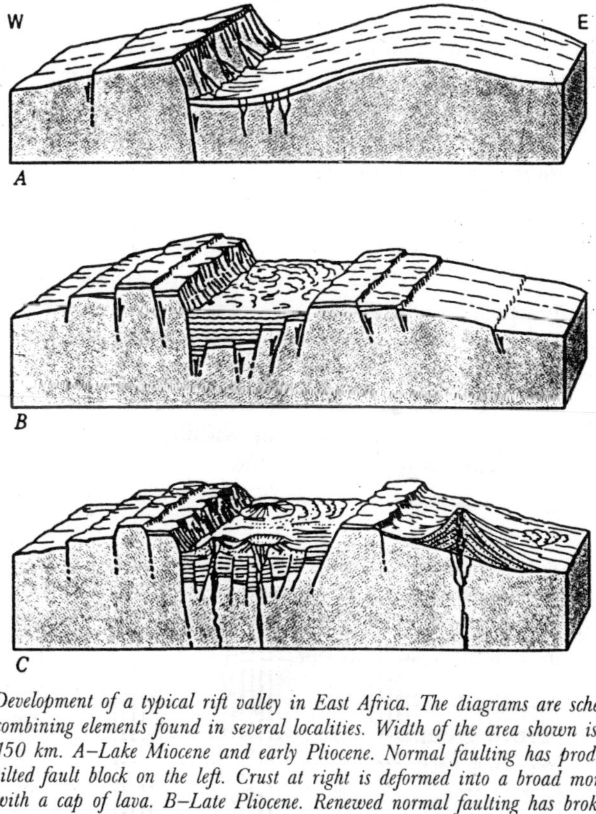

Fig. 1.8. Development of a typical rift valley in East Africa. The diagrams are schematic, combining elements found in several localities. Width of the area shown is about 150 km. A–Lake Miocene and early Pliocene. Normal faulting has produced a tilted fault block on the left. Crust at right is deformed into a broad monocline with a cap of lava. B–Late Pliocene. Renewed normal faulting has broken the valley floor into narrow blocks and raised the eastern side. The rift valley is now a graben structure. Lava flows have filled the valley floor. C–Pleistocene and Holocene. After another episode of minor faulting, extrusive activity has built volcanoes in the rift valley and on the flank of the uplift.

different species of cichlids exist per 10 square meters (Reinthall & Lewis 1986). The fish have adopted a large number of specialized feeding behaviours. For example algae and diatoms grow in the water column, on rocks and sand; different species have specialized their feeding behaviour to live in each of these conditions.

In addition, some cichlids feed on the scales and fins of other fish. Lake Malawi is anoxic at about 230 m, and bacteria are the dominant organisms in anoxic waters. In addition to nutrients sinking to depth and being lost, another way in which there is a loss of nutrients from the lake is through the hatching of midges (Chaoborus and Chironomus). So many midges hover in the air at one time that they from dark clouds above the surface of the lake. When the midges hatch and blow onto the land they carry valuable nutrients from the lake. As in the case of the finches and turtles which Darwin found in the Galapagos Islands, the many related fish species existing within Lake Malawi raises questions regarding the development of new species.

Wetland Communities

One common successional hypothesis treats open bodies of water as ephemeral features of the landscape which over time fill in and disappear. Starting with an open body of water, there is a submergent stage (submergent water plants become dominant), an emergent stage (shallow water herbaceous plants dominate) and then "climaxes" with a closed stage of either a wet prairie or a wet forest. There are situations, such as within the floodplain, where this sequence may be appropriate. However, because of the qualities of the organic materials that have filled the depression, organisms may continue to play a role in maintenance of the body of water. There are a number of wetland communities of plants and animals which thrive within and are an inextricable component of the drainage system. A few of these communities, the bog, marsh, and swamp are briefly described below.

Bogs

Two broad groups of soils are commonly recognized: mineral and organic soils. Organic soils are composed of large accumulations of mosses or sedges, and are commonly referred to as peat. They have the highest organic matter content of any soil. One of the outstanding physical properties of these material is their high water holding capacity. Minerals soils hold about 20 percent to 40 percent of their weight as water while undecayed moss or sedge peat can hold 12 to 20 times their dry weight as water. These materials are veritable sponges in the landscape. Organic deposits of peat occur in many places on the Earth. They are located in structural basins, such as the Everglades of Florida, which has sedge derived peat, but the major location are in glaciated landscapes. Extensive deposits occur in northwestern Eurasia and North America. Sphagnum moss composes the peat of the bogs of large areas of North America and Eurasia.

The water in these bogs is cold, anaerobic, high in dissolved organic matter, and low in minerals. The moss is essentially sterile, containing neither bacteria nor the organisms of decay that are usually associated with soil. The large bogs exert much influence on the hydrology of the surrounding regions. They may cause either an increase or decrease on the annual spring flood peak, depending on how the bog is tied into the groundwater system. In general, the bogs reduce storm peaks significantly and augment streamflow during dry periods. Thus, rivers that are supplied by the bogs tend to have a more constant flow than other streams (Novitzki 1979). Bog communities thus exert a moderating influence on the stream system.

Marshes and Swamps

Marshes and swamps develop in moist climates, generally in low lying depressions, and play an important role in channelling the runoff of a stream system. If the depression is shallow and poorly drained, and supports plants and emerge above the water surface, it qualities as a marsh or swamp. When water enters the depression, the velocity decreases rapidly, and the organic matter, sedimentary particles, and dissolved minerals that have been carried by the stream flocculate and accumulate within a relatively small area. These small areas are the recipients of the nutrients and organic wealth of the watershed and are exceptionally rich places on the earth.

Marshes and swamps are distinguished from each other by their vegetation. Vegetation varies with climate. The seasonal, continental mid-latitude climates support cattails and reeds that make up the marsh; warmer climates support the growth of trees such as cypress which make up the swamp vegetation. Emergent marshes are extremely productive environments. They annually produce between 450 and 2,320 g/m^2 (dry weight) of above-ground material and compare favourably with saltwater marshes in their productivity

(Good *et al*,. 1981). In order of their productivity are cattails (Typha), reed canary-grass (Phalaris arundinacea), reeds (Phargmites communis), sedges (Carex), and bulrush (Scirpus).

Marshes can undergo enormous fluctuation of productivity from year to year. Prairie marshes have been described which vary by as much as 18-fold over a period of 5-20 years (Good et al., 1978). Marsh variability is related to two major factors. First, weather fluctuation leads to changes in water depth that strongly influence the plants which can be supported within a body of water. During wet years high water may flood out emergent plants, while dry years may result in the filling in of a marsh with emergent plants. Severe drought may even lead to the drying up of a prairie pothole, when the unprotected bottom may become deepened by wind erosion. In North America an animal plays a major role in marsh vegetation, the muskrat (Ondatra zibethica).

Muskrats are rodents that have an impressive capacity to reproduce. They have six or more young per liter and under optimal conditions, can have four litter per year. Muskrat population have been reported to grow so rapidly that they literally "eat out" the plants from the marsh (Weller 1981). The muskrats use the plants for food and construction. The rootstocks, rhizomes, and tips of the plants are eaten and the stems are used to construct a platform and left to rot. In the early stages of reestablishment in a marsh, muskrats tend to construct lodges nearly 2 m high and 4.5 m in diameter. However, the lodges are half that size when the muskrat population has removed a large proportion of the plant material.

Reproduction rates have also been reported to undergo changes that reflect the conditions of the marsh. Thus, as the marsh becomes crowded, lodges decrease in size and the number of surviving young decrease. These dynamic activities affect the numbers and species of birds which inhabit a marsh. The greatest number of individual birds are supported when the whole marsh is dominated by a thick growth of emergent plants. However, when there is a roughly equal balance between open water and emergent vegetation, then the maximum number of species can be supported by the marsh. When muskrats open up the vegetation around their lodges, they create habitat diversity that attracts an increased diversity of bird species (Weller 1981). The most extensive swamps in North America occur along the Atlantic and Gulf coastal plains.

The most notable examples include the great Dismal Swamp, Okefenokee Swamps, and Everglades. The Great Dismal Swamp consists of nearly 200 km^2 within the Atlantic coastal plain. The swamp is believed to be at least 80,000 years old (Kirk 1979). The oldest pollen is from pines and spruce. The modern vegetation consists of a variety of tree species including cypress, white cedar, and tupelo gum (Nyssa aquatica). The swamp acts to stabilize the groundwater table in the region. During periods of very heavy rains, the swamps controls floodwater by temporarily storing excess runoff and releasing it slowly as runoff or into the groundwater system. The area abounds with an amazing variety of plants and animals, some of which are poorly known (Kirk 1979). Within the Everglades of South Florida and the Gulf Coastal swamps, one organism has had an especially important role within the environments system, the American alligator (Alligator mississippiensis). Alligators are active diggers. They dig out channels and build large mounds for nest sites. They also dig deep pools or "gator holes" in the swamp.

During periods of extreme drought these pools provide the last remaining inhabitable sites in the swamp for birds and other animals to survive. When the drought breaks repopulation of the swamp takes place from the animals which have survived at gator holes. The nesting mounds of alligators contribute to the variety of the swamp environment by providing nest sites for several bird species, such as herons and egrets. Thus, marshes and swamps are complicated habitats that depend on energy expenditure and behavioural adjustment of specialized water-dwelling animals inhabitants. The activities of these animals create conditions that provide opportunities for existence of the larger plant and animals community.

Beaver Impoundments

Beavers are a sine qua non of the influence of animals on streams. They have an unusually large impact on stream systems and the development of valley soils and vegetation of the Northern Hemisphere. Two species are recognized, the Eurasian beaver, Castor fibre, and the North American beaver C. canadensis. The beaver is a large rodent, second in size to the capybara (Hydrochoerus hydrochaeris), a relative of the guinea-pig and inhabitant of tropical America. The beaver attain a length of between 90 and 120 cm and a weight of up to 30 kg. The present range of the beaver is only a fraction of its former range. In the Old World the range of

C. fibre once extended from Scandinavia to Spain and from Britain to Siberia. Now there are only limited population in the lower Rhone, the middle Elbe, and the southern part of Norway. The former range of C. canadensis include the northern half of North America with extensions into the southern Appalachians and the southern Rocky Mountains. They numbered in the thousands of millions and were and attraction to early traders who sought their pelts.

In general beavers, were hunted for food (because of the scales on their tails they were defined as fish and could be consumed by Catholics on Friday and during Lent), for fur (they have magnificent coats which are durable and easily tailored) for castoreum, a scent gland substance which was highly valued for it medicinal properties during medieval times. For these reasons and because they occupied flood plain soils that humans valued for agriculture, the populations of beavers were cut to a fraction of their former numbers by hunters and trappers. But, many soil and vegetation features can be understood only by considering the influence that beavers have on river valleys. The activities of the beaver are entirely bound up with trees, flowing water, and the construction of dams.

Beavers create deep ponds or maintain water levels in preexisting depressions by building dams across streams. These are usually made of logs and stick, although dams made of boulders have also been described (Chapman and Chapman 1937). The impoundment stores water and serves the beaver by providing a site for a lodge which is safer from predators. The food of the beaver consists of the bark and shoots of poplar, willow, alder, elm or oak plus vegetation which is favoured by the expanded riparian environment that the beaver pond creates.

The beaver obtains food by chopping down trees. The bark is stripped and the inedible portions of the trees are moved to the river for the construction of the beaver dam or the lodge which is built in the impounded waters behind the dam. In order to move large logs, beavers construct canals. Castor canadensis is acknowledged as the superior dam builder. The largest recorded dam built by C. fiber was found in the Voronezh region of Russia. It was 120 m long, 1 m high, and 60-100 cm wide (Frisch 1974). One of the largest structures constructed by C. canadensis was on the Jefferson River in Montana; it was 700 m long and could bear the weight of horses and then riders (French 1974). The largest dams were probably maintained by generations of beavers for centuries.

Beavers regulate the water level of the beaver pond by constantly adjusting the dam. During high water, the dam can be adjusted by lowering the crown. During low water the beavers tighten up the dam to reduce seepage. The splendid teamwork of beavers building and maintaining a single dam is itself magnificent. A series of dams within a number of tributaries of a drainage basin operate together to create an impact on the landscape. The concerted efforts of many beavers must have changed many valleys of he Northern Hemisphere during the Holocene from relatively physically and biologically homogenous habitats into heterogeneous environments supporting diverse groups of organisms.

The process can be imagined within the high valleys of the Rocky Mountains. Prior to the movement of beavers into high valleys, conifers probably dominated the landscape. As beavers move in and construct dams, the impounded waters flood out the conifers. Each pond creates a site which is attractive to migrating waterfowl and other birds. The riparian environment of the shores of the pond supports new types of plants (poplar, willow, alder), that supply the food for the beaver population. The pond checks the movement of sediments down stream and these particles and to the richness for the aquatic organisms of the pond. The enlarged surface area of the enriched pool of water supports high levels of photosynthesis and releases increased amounts of organic carbon that in turn enhances the food supply for aquatic organisms down stream.

Eventually, the impoundment becomes filled with sediments, and beavers abandon the site. As the beaver pond drains it becomes a thick lens of rich humus that supports the rapid growth of grasses and forbs (non-grass species) and establishes the beaver meadow. These soils have high water-holding capacity and slowly release water to the water table and the stream. The activity of the beaver becomes the primary force in the evolution of the soils in the riparian zone. They influence the diversity of habitats and establish structures which argument the slow but steady movement of water through the drainage basin. These feats give the beaver its reputation as the unparalleled engineer and builder of the animal kingdom.

The Estuary

Eventually, most freshwater stream flow to ocean level and meet the sea. Under special conditions, the flow of freshwater to the ocean is partially enclosed and the outlet to the sea restricted to a relatively narrow opening. These places are referred to as estuaries. The most important estuaries include drowned river

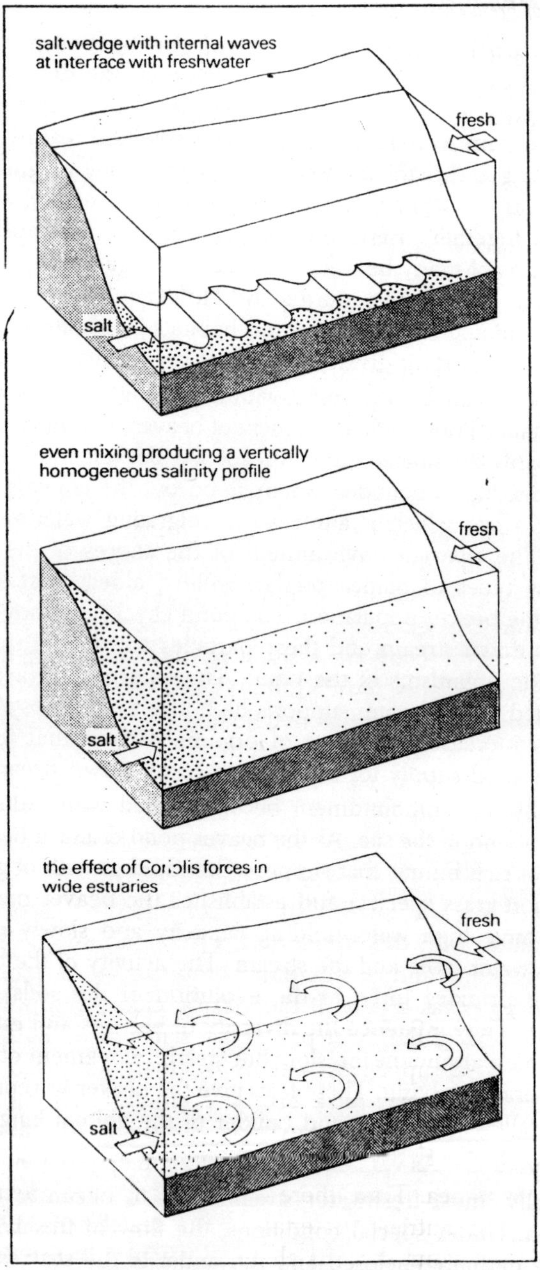

Fig. 1.9. Mixing of water masses in estuaries.

valleys, such as Chesapeake Bay; down-faulted sections of the Earth's crust, such as San Francisco Bay; drowned glacial troughs, or fjords, found in Norway, Alaska, New Zealand, and other glaciated mountainous coastal regions: and saltmarshes and lagoons that are formed behind barrier beaches. Within the estuary, fresh water from the land flows upon and meets the denser salt water of the ocean. The interface between the two different types of water is nutrient-rich and highly productive site for organisms. But these sites are also physically demanding and only a few species can tolerate the rigors of successive exposure to fresh and sea water, and the wide fluctuation in temperature associated with these environments. One component of the estuary is shallow saltmarshes that play an important role for organisms. Saltmarshes are regularly rinsed by the ebb and flow or the tides.

Waste products are diluted and removed and nutrients are brought in from the land and the sea. The materials of streams flocculate as the mixture of clay size particles and algae meets the sea water. They settle out, leaving the water clear (Avnimelech et.al. 1982). In the saltmarsh, plants need not expand energy in collecting minerals. They are assisted in obtaining minerals from many organisms. Clams, oysters, and mussels trap nutrients as they feed, and deposit the wastes as pseudo-feces. In these nutrient-rich waters, plants grow, die, and cycle rapidly as they are consumed by the inhabitants of estuary. The base of the food chain of the estuary is marsh grass or Spartina. The major energy flow is by way of the detritus food chain rather than the grazing food chain.

In estuaries along the coast of Georgia, U.S.A., for example, Spartina alterniflora dominates, and organic detritus (over 90% of which originates with the growth of Spartina) is the link between primary production and consumption. Only a small portion of the grass is consumed as a part of the grazing chain (Ranwell 1972). Estuaries are one of the most productive habitats on Earth. They provide huge food supplies as well as sanctuary for many species of waterfowl. They serve as a breeding ground and nursery for many marine fish species. They store nutrients for larval and juvenile marine life. As various species mature, they leave the estuary and head out to sea. Thus, there is a constant movement of the rich organic and nutrient materials to the sea (Honjo 1982). As tides move in and out there is also a movement of the materials out to sea. And thus it is that the stream flows to the sea and enriches the shallow coastal waters of the marine environment.

2

RESOURCES OF WATER

Human use of land depends on the availability of water, and the Jack of water is a limiting factor in many regions. Where water resources are relatively abundant, this constraint will seldom be confronted, and then only during extended periods of drought. Where water is not so abundant, this constraint is confronted all the time, so the necessity for water management is recognized by most people in such regions. One should not conclude from this that water-resource management is only a concern in dry areas. It is not. Although water resource management is such an integral part of the functioning of any society that it may be taken for granted, it is always a matter of professional concern.

The Hydrologic Cycle

Every examination of water resources begins with an examination of the hydrologic cycle within the region of interest. The general features of this cycle are represented pictorially which shows the pathways followed by water initially falling on the surface as precipitation. Some of this precipitation returns to the atmosphere through evaporation and plant transpiration. This part of the cycle is referred to as evapotranspiration. Some of the precipitation also moves overland as runoff to streams, lakes, and in some areas, the sea. And finally, some of the precipitation moves down into the ground through the zone of aeration, where the soil is not saturated with water. This infiltrating water may reach the level where the soil or rock is saturated with water. This is known as the zone of saturation, the upper surface of which is called the water table.

Water reaching this level becomes part of the ground water. Each of these elements is linked to the others. A surface-water body such as a stream is fed by direct precipitation, by surface runoff, and by groundwater flow.

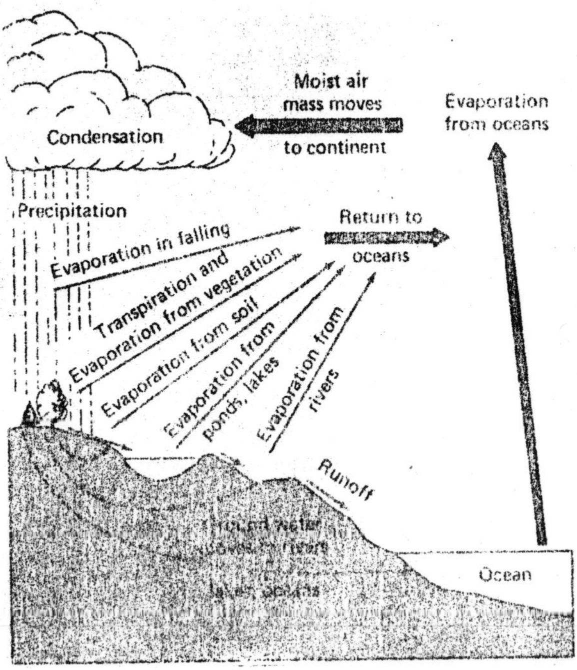

Fig. 2.1. The hydrologic cycle traces the movement of water through the atmosphere and oceans and its flow over and beneath the land surfaces.

The atmosphere is fed by evaporation from the ground surface, rivers, lakes, and the ocean, and by plant transpiration as well. A major task of the hydrologist is to measure the flow of water from one point in the hydrologic cycle to another. These data must be gathered over time since these flows all vary with time. The analysis and use of hydrologic data showing vacillation with time are central to hydrology. Water flow is viewed in the framework of a hydrologic system consisting of many separate elements linked to one another by pathways along which water moves. A diagrammatic representation shows more clearly than the pictorial representation the different pathways that water can follow to arrive at anyone destination.

Hydrologic information comes from studies by meteorologist s and hydrologists. In the United States, the collection and recording of these data are primarily the responsibility of the U.S. Geological Survey Water Resources Branch and the U.S. National Weather Service, although many government agencies and other institutions also help. These data are published in USGS water-supply papers, hydrologic atlases, and other series prepared by the Water Resources Branch. The USGS also publishes the National Water Summary, which identifies hydrologic events and related issues in all 50 states (U.S. Geological Survey, 1984). We will first consider how these measurements are made. Then we will examine some general information on water use. Next we will look at issues of surface-water management. Finally, we will look at issues of groundwater use. Although this division is convenient, there are many instances in which we must consider the relationship between ground water and surface water because the two are closely connected.

Measurements Employed in Hydrology

Here we consider some aspects of measurement and the units that are employed by hydrologists in reporting these measurements. We can think of the flow of water from one element in the hydrologic system to another as if it took place in a pipe connecting the two elements. Water enters this "pipe" from one element, moves through the pipe at some rate, and is discharged from the pipe into the next element. A simple example is represented by two ponds connected by a river. The river represents the pipe in which water is transported from the upper pond to the lower. The discharge into the lower pond is measured in terms of the volume of water released in a given time. The discharge from one end of a pipe may be expressed in units of L^3/t, where L is a unit of length and t is a unit of time. Common units used to express discharge are cubic feet per second (ft^3/sec), generally written as cfs, and cubic metres per second, generally written as m^3/sec. In the United States, hydrologists also use an additional unit, the acre-foot. An acre foot is the volume of water that would result if an area of one acre was to be covered by 1 foot of water. In general, instantaneous discharge is expressed as the product of the cross-sectional area (A) of the "pipe" and the mean velocity of flow of water through the "pipe" at the point of discharge.

Measuring instantaneous discharge requires measuring the cross-sectional area and the average velocity of flow. **It** is typical

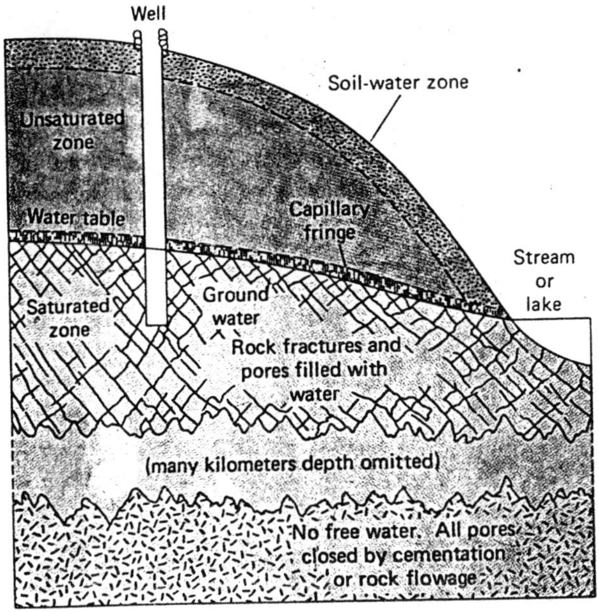

Fig. 2.2. A schematic cross section through the various zones of subsurface water.

relationships in streams in regions that are not arid. Note that the cross-sectional area of the water-filled channel increases downstream. Because of this, and because stream velocity generally increases downstream, the volume of water passing an upstream point at a particular moment usually is less than the volume of water passing any point further downstream at the same moment. The figure also shows how the areas of the water-filled cross-sections increase during periods of increased precipitation. Measurements of discharge are made at locations called gaging stations, where the cross-sectional area of the water filled part of the channel can be measured directly. At such a station, the mean velocity of flow of water past the station may be determined by measuring the velocity at several different points and depths and then averaging these by appropriate procedures. The cross-sectional area is designated as A and the mean velocity as v. The instantaneous discharge, designated by the letter symbol Q, is then the product of the measured values of A and v at the time of measurement:

$$Q = Av$$

It is apparent that the cross-sectional area of the water-filled channel increases as the water level in the river rises, and it decreases as the water level falls. Thus rising water level leads to increasing discharge; and falling water level leads to decreasing discharge. As the water level rises, the mean velocity of flow also increases, which contributes to increased discharge. A record of the fluctuation of discharge with time is called a discharge hydrograph. The U.S. Geological Survey publishes hydrographic information of this kind for thousands of rivers in the United States in its water supply paper series. The discharge values at each gaging station are measured by continuously monitoring the water level or stage of the river.

The measured value of stage is converted to a value of discharge by the use of calibration curves, called stage-discharge or rating curves, which are based on a study of the relationship between the cross-sectional area (A) and stage, and between the mean velocity (v) and stage, at each gaging station. Once these calibration curves have been prepared, it is no longer necessary to measure cross-sectional area or stream velocity. The simple measurement of stage is sufficient. Another important measurement determines the volume of water in storage in a reservoir or in a segment of a stream channel. It is easy to picture the water in storage in a reservoir. If we know the configuration of the bottom of the reservoir, we can determine the volume of water in that reservoir for any observed water level. The same concept applies to a river channel, even though water is moving through that channel, and it is perfectly proper to speak of water in storage in a channel. Finally, we are often interested in changes in storage during time intervals of observation, especially in monitoring water supplies in reservoirs. These changes in storage in a reservoir. If we know the configuration

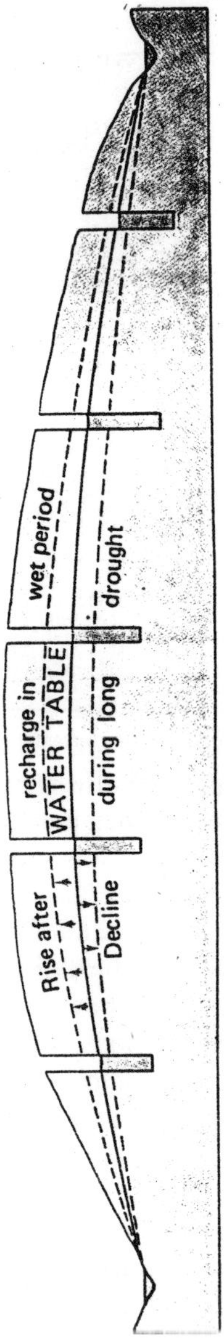

Fig. 2.3. Position of the water table can be determined by the height of standing water in wells.

of the bottom of the reservoir, (ΔS) per unit time is equal to the difference between the average rate of inflow (I) to the reservoir and the sum of the rate of outflow (O) and average evaporation (E) during that time. For finite periods of time, the equation is as follows, where Δt is the time period of interest:

$$I\Delta t = (O\Delta t + E\Delta t) + \Delta S$$

With this basic terminology in hand, we can examine the subjects of water use and surface-water management.

Water Use

Every region is characterized by its own natural hydrologic cycle, which existed before the region was settled or developed. Humans occupying that region use water moving through various elements of the local hydrologic system for agriculture, livestock, industry, power development, natural resource development, and human consumption. Much of the water drawn from one element of a local hydrologic system is not returned to it. For example, much of the water withdrawn for irrigation is lost to the atmosphere through evapotranspiration. Such a use is called a consumptive use. Many other uses allow almost all of the water to be returned to the same element from which it was withdrawn. For example, river water diverted through an electrical power generating plant is passed through the power-plant turbines and returned to the river. Such a use is called a nonconsumptive use. Many other uses allow for the return of part of the water withdrawn. For each such partially consumptive use, we can describe the fraction lost in terms of net consumptive use. These terms and their application are explained fully in USGS Circular 1001 (U.S. Geological Survey, 1983a). The water returned to an element of the surface-water system helps to maintain the volume of water in that element.

However, this returned water is likely to be warmer than, or chemically different from, the water that was withdrawn. As a result, this returned water is likely to change the water quality in that element. Ordinarily, such changes in water quality are seen as adverse, and are therefore described as causing a decline in water quality. The distribution of water among users in the United States may be seen from a diagram the distribution and changes in distribution of water use in the United States. In the United States in 1980, the average rate of water withdrawal was 450 bgd (billion gallons per day). Of this total, 150 bgd went to irrigation and 255 bgd to industry.

Most of the water used by industry was for cooling thermoelectric power plants (210 bgd). Of all the water withdrawn each day, about two thirds were returned to surface-water bodies. The other third was returned to the atmosphere as water vapour or was incorporated in manufactured products. Irrigation was the dominant consumptive use of water, accounting for 83 bgd in 1980. The sources of the 450 bgd were surface water (64 percent), ground water (20 percent), and saline waters. (16 percent). From our point of view, these national patterns of water use are of less interest than patterns of water use within a single area or region. The reason for this is simple–in most regions, the water used comes from supplies within that region. Where this is true, issues of water-resource management are initially local or regional issues, not national issues. Thus, if we want to learn how scientific information is employed in dealing with water-resource issues, we need to learn about local water-use patterns and supply problems. In a particular region, water-use patterns evolve by accident, design, or some combination of accident and design. And once a pattern is created, it is not easily broken.

As a consequence, we seek to ensure that the existing water-use pattern can be maintained and even expanded. To do this usually requires an intricate interplay between scientific knowledge, technological manipulation, and a body of water law and custom. Where water-resource problems arise, they will be dealt with, if at all, through the use of scientific information tightly constrained by the legal system and political considerations. We can see this by examining the hydrologic features of the Colorado River basin and the water use in that basin. This provides a case history and an introduction to basic concepts that must be dealt with in any examination of surface-water management.

Case History: Surface-water Management in the Colorado River Basin

Geography and History

The Colorado River basin is a 609,520 km2 area, within which surface runoff and ground water move toward the Colorado River and its tributaries. The basin is divided by law (the Colorado River Compact of 1922 into upper and lower basins. The entire basin lies wit bin an arid region. Average annual precipitation in the upper basin is about 16 inches and in the lower basin is about 8 inches. Most of this precipitation is lost to the atmosphere through evapotranspiration. In the upper basin these losses are estimated to be 85 percent and in the lower basin, even more than this. Yet the upper basin supplies 83 percent of all the water that flows down the Colorado River. The upper basin supply is obviously of critical importance to all users of Colorado River water. Serious consideration of the use of scientific information in water-use management in the United States began with the work of John Wesley Powell in the Colorado River basin.

In 1869, Powell explored the Colorado from the mouth of the Green River in Wyoming to the Gulf of California. Later he prepared a major study of the arid regions of the United States, (Powell, 1878), in which he presented his case for water-use management in these regions. As the second Director of the U.S. Geological Survey, he made strenuous efforts to convince the U.S. Congress of the need to create a scientific data base for the Colorado River Basin. Powell's scientific work is important because it led him and others to the view that land-use development of the region should take such factors as water availability into account. His personal history (Rabbitt, 1969) is of interest as well, because it portrays the conflicts that can arise between a scientist and his opponents. These conflicts are obvious in all efforts to manage the Colorado River.

Basic Hydrologic Information for the Colorado River

The hydrologic system of the Colorado River Basin functioned in its natural state before the basin was occupied by European settlers. The flow in the Colorado River under such natural conditions is referred to as the virgin flow, and the term virgin flow is used today to describe the flow that would occur in a river if there were no human influences. All efforts to manage water withdrawals and to design systems of dams depend on the availability of a record of natural flow–also called virgin flow–in the river a record that will show the short- and long-term variations in flow caused by natural phenomena.

From this hydrographic record we can obtain such important additional information as the annual high and low flows, the total annual flow in each year, and the mean annual flow over the period of record. For the lower Colorado River, such information is obtained at Lees Ferry, Arizona, where a USGS gaging station is located. Flow in the Colorado River has been monitored at Lees

Ferry since 1921, and flow at some upstream points has been monitored even longer. This record is one of the best bases on which to formulate predictions of the range in scale of future flows in the Colorado River. The flow record the measured annual flow at Lees Ferry, called actual flow, and the estimated virgin annual flow.

The actual flow is the total volume of water measured in millions of acre-feet (mar) passing through the river in a 12-month period. If no water had been withdrawn upstream, then this actual flow would be the same as the virgin flow. Since upriver diversions have ranged between 2 and 4 maf each year, the observed flow has generally been 2 to 4 maf less than the estimated virgin flow. Some of the important hydrologic facts that bear upon water-management decisions. We see first of all that the virgin annual flow during the period of record has varied strikingly from a low of 5 maf to a high of 24 maf. The overall mean annual virgin flow computed from the entire record is 14.9 maf. However, if the record is divided into two parts, we find that the mean annual virgin flow at Lees Ferry for the 35-year period 1896-1930 was approximately 17 maf; and for the 35-year period 1931-1965, it was approximately 13 maf. This basic record leads us to ask some questions:

1. Has there been a long-term decrease in the virgin flow of the Colorado River?
2. Are the estimates for the earlier 35-year period less reliable than the estimates for the more recent period? If they are, then we cannot conclude that there has been a long-term decrease.
3. Do years of low flow or high flow tend to occur in groups.

All of these questions have been addressed repeatedly by analysts (see National Research Council, 1968, 1977, 1981) because the answers are of critical importance in assessing past and future plans for water management. The scientific information considered in the preceding paragraphs was obtained by direct measurement during a relatively short period of historic record. Efforts have been made to expand this record and to put it into perspective by using tree-ring chronologies for a 400-year period back to 1564 A.D. (Stockton, 1977). Studies of tree rings allow estimates to be made of precipitation in the basin; and these estimates serve, in turn, as the basis for estimating virgin flow in the Colorado River.

A sample of this record suggests that there have been periods of tens of years in which the virgin flow was well below 13 maf, and that there have been individual years in which the virgin flow was below the historic low flow recorded in 1934. This information, which seems to be so basic, takes on considerable importance when we examine the pattern of past use and the plans for future use of water from the Colorado River Basin.

Water Withdrawals and Water Management Problems

The water-management problem, simply stated, is to determine how water supply and water use can be managed so that existing patterns of water use are served, and where new patterns are proposed, to find new schemes of accommodating them. The water-management problem began to take shape in the Colorado River Basin during the years in which Powell was active as a U.S. Geological Survey scientist. Development of the problem is reviewed in many sources. Here the principal sources are the USGS Professional Papers 441 (lorns, Hembree, and Oakland, 1965) and 486 (Hely, 1969; McDonald and Loeltz, 1976), the National Research Council studies 1968, 1977, 1981), the California Water Atlas (Kahrl, 1978) and the historical review of Rabbitt (1969).

The problem was created by the successive diversions of water from the river for consumptive uses, which were largely agricultural; this has resulted in progressive diminution of river flow. The first such diversions took place in 1840 in the northern part of the basin, when a Mormon settlement in Wyoming diverted water from the Green River, a major tributary to the Colorado, to provide water for irrigation. By 1880, the Uncompaghre Valley had been put under irrigation; and in 1909, the U.S.. Bureau of Reclamation constructed the first diversion tunnel, a 6-mile long tunnel carrying water from the Gunnison River to the uncompaghre. Diversions in the southern part of the basin were started at a bout the same time

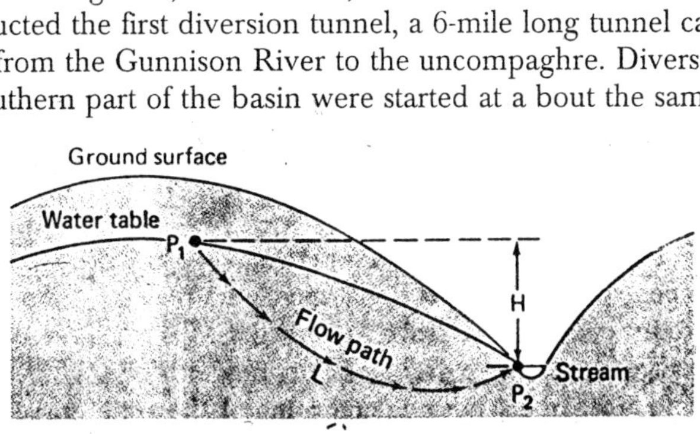

Fig. 2.4. Diagram showing the meaning of terms used in Darcy's law.

as those in the north, and they soon came to dominate the pattern of consumptive use. Therefore the focus here is on the lower Colorado, the segment down-river from Lees Ferry, Arizona, in the southern part of the basin.

The first diversions in the southern part began in 1877 when Colorado River water was diverted to the Palo Verde Valley in California. In the 1890s, the Yuma Valley in Arizona was placed under irrigation. Then, in 1901, the Imperial Valley, in California, which had been called the Imperial Desert, began to receive water fed from the Colorado via the Alamo channel, which had been cut through Mexico for a distance of 80 km. Development of the Imperial Valley represented the first truely large-scale diversion from the Colorado River. In 1903 and 1904, the Alamo channel was blocked by deposits of silt.

A new channel was made, and the channel cut became enlarged in 1905 during a period of floods on the river. This new channel became large enough to allow all of the Colorado River water to flow into the Salton Sink, which thereby became the Salton Sea. Early diversions for agricultural use were later augmented by new diversions for metropolitan water-supply purposes. These were begun in 1941 with completion of the Colorado River aqueduct, which carried water Lake Havasu, created by the Parker Dam, to Los Angeles. By 1961, this aqueduct was serving the water-use needs of 7,500,000 people. These and the many other diversions are, which shows the dams below Lees Ferry and the major diversions. Additional diversions are planned under the authority of the Colorado River Basin Project Act of 1968; which included authorization for the Central Arizona Project. This project is intended to supply 1.2 maf of water to the Phoenix and Tucson areas when it is completed. These diversions may seriously deplete flow in the river. The reservoirs created to store water for diversion will also increase evaporation, further depleting flow.

The Framework of Colorado River Basin Water Law

The water-use management system for the Colorado River has evolved under the influence of a set of legal controls created by the U.S. Congress and the Supreme Court and under the influence of the basic principles of water-resource law in the western United States. The key principle in western water resource law is represented by the doctrine of prior appropriation, which says that a user who has once established a pattern of beneficial use of previously unappropriated surface water has, in so doing, established a future right to continued water use on the scale initially established. Thus, the early use of Colorado River water in the Imperial Valley represents the establishment of a pattern of beneficial use of previously unappropriated water. The legal controls established by the U.S. Congress and the Supreme Court constitute a preface to our examination of the application of scientific findings about the Colorado. The more important controls include:

1. The Colorado River Compact of 1922,
2. the Boulder Canyon Project of 1928,
3. the Rio Grande, Colorado, and Tijuana Treaty of 1944,
4. the Decree of the U.S. Supreme Court, 1964, Re: Arizona v. California, Colorado River, and
5. the Colorado River Basin Project Act of 1968.

These are reviewed in Hely (1969), Kahrl (1978), and the National Research Council (1981). Here we need only sample them to see the problems ahead. The Colorado River Compact of 1922 apportioned water in the Colorado River system to an "upper basin" and a "lower basin", and it also provided for the satisfaction of any water rights of Mexico that the United States might recognize at any future time.

One of the most important provisions of the compact is that the upper and lower basins are each allowed to allocate 7.5 million acre-feet of water per

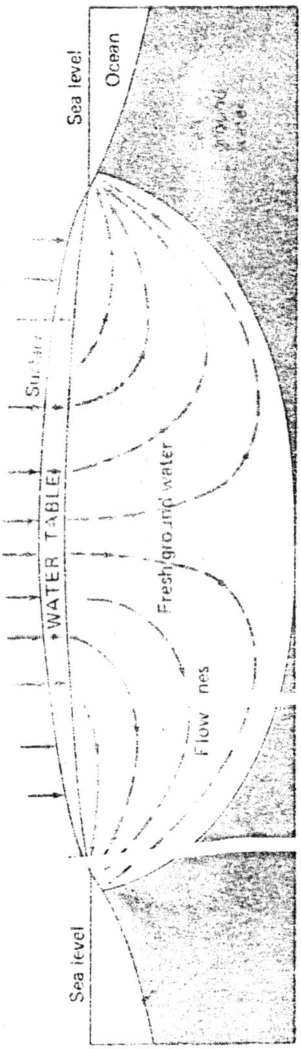

Fig. 2.5. Relation of fresh to salt ground water under an island or peninsula.

year to consumptive use. Another provision is that the Upper basin must ensure that 75 million acre feet of water are allowed to flow past the measuring point (called the compact point) as Lees Ferry, Arizona, during any to-year period. This provision reflects the knowledge that most of the flow in the lower Colorado River is derived from the upper Colorado River. The compact did not allocate water among competing users. The Rio Grande, Colorado, and Tijuana Treaty awarded 1.5 maf to Mexico, placing an additional demand on Colorado River water.

Allocation of water among competing users is represented by the decree of the U.S. Supreme Court of 9 March 1964, which specified that "of such 7,500,000 acre-feet of consumptive use. there shall be apportioned 2,800,000 acre feet for use in Arizona, 4,400,000 acre-feet for use in California, and 300,000 acre-feet for use in Nevada." If less than 7,500,000, acre-feet are available in a given year, then it is the responsibility of the U.S. Department of Interior to apportion the use, among the three states. This provision may be put to a test when the Central Arizona Project is completed since the 1968 legislation authorizing this project requires that California first be assured of receiving 4.4 maf per year before any water is diverted into the Central Arizona Project system.

We see from these provisions that water is allocated by law. These allocations reflect existing patterns of development and use. Can these allocations be met in each and every year in the future? If they cannot, then major economic and political problems will be faced in those years in which the allocations cannot be met. The natural and social scientists who have studied the Colorado River system have the task of assembling factual information about the present system and developing information about the history of flow in the system. Another, goal is to predict the future behaviour of the system.

Scientific Information and Water Management

The Colorado River, like most large rivers, is a managed river. It is managed to provide water for irrigation, power generation, municipal use, and recreation. We have seen that the first Stages in the evolution of this water-use management system were taken before there was much scientific information available. Even in 1922, when the first legal controls were being placed on the system, decisions were made to allocate water in the belief that at least 14.8 maf were available for distribution in an average water year. With this compact as a framework, a water-management system was developed. One of the major features of the water-management system is a series of dams that created nine reservoirs along the Colorado. These reservoirs have a total capacity of 61.6. maf. However, no more than 56.4 maf are supposed to be stored in them, so that 5.2 maf of additional storage capacity will be available for use when there is a threat of flooding. With this system, water can be stored during periods of high 'flow and then can be released from the control points. These released are intended to provide for the diversions prescribed by law.

The reservoirs can provide the water needed to meet the legal requirements during periods of low flow. This same system is also designed to provide for electrical power generation and flood control in the lower Colorado. Inevitably, these functions can come into conflict. If the reservoirs are kept filled to meet water demands during dry periods, they are unable to provide needed flood-control storage during periods of high flow. This situation was faced in 1983, when severe flooding was experienced in the lower part of the Colorado River Basin. Scientific management of surface-water resources is concerned with finding the best possible compromise solutions to meeting conflicting demands such as these (see National Research Council, 1982). Unfortunately, it now appears that the system may not be able to supply all the water needed for uses that now exist, not to mention those that are being planned for the future.

In other words, scientific realities for the Colorado may place constraints on the exercise of legally assured rights and on land and resource-use developments being considered. The major part of virgin flow in the Colorado comes from the upper basin, and the future of the entire Colorado River Basin therefore depends on the magnitude of that flow, on withdrawals within the upper basin, and on the legal requirement that the upper basin deliver at least 7.5 maf per year to the lower basin. Let us look at these figures. In 1974, water consumption in the tipper basin was estimated' to be 3.7 maf, divided among various uses.

The Bureau or Reclamation has estimated that at least 5.8 maf should be available for use within the upper basin in any year. The Colorado River Compact actually specifies that 7.5 maf are to be available for use in the upper basin by legal right. However, the record of virgin flow at Lees Ferry suggests reasons for concern.

For example, if average annual flow at Less Ferry in the next 25 years is only 13.1 maf, which was the average annual flow from 1931 to 1974, then the maximum amount of water available for consumptive use in the upper basin would only be 4.9 maf per year (13.1 maf–8.2 maf=4.9 maf). The figure of 8.25 maf is used by Dracup (1977) instead of the 7.5 maf to allow for commitments in addition to those in the 1922 Compact. Use of the 7.5 maf figure would, of course, indicate that more water would be available for consumptive use in the upper basin.

If average flow during a period of 5 to 10 years were to be only 10 maf per year, as it has been in the past, then the maximum amount of water available to the upper basin would only be 1.75 mar per year (10 maf–8.25 maf=1.75 maf). This is about half of the volume actually used in 1974. Clearly, in an unmanaged river, there is no way in which the existing commitments and current consumptive uses could be met in the event of a period of low flow. As we have seen, the Colorado is managed by control of flow at the various dams. Let us look at the way in which this managed system might function during a period of relative drought, following Dracup's example.

The major dam that provides significant control in the upper basin is the Glen Canyon dam just above Lees Ferry, which is designed to provide a usable storage capacity of 25 maf in Lake Powell. The total upper-basin storage capacity is 31 maf. This stored water is intended for use during periods of relatively low flow. Suppose that total flow in a to-year period was 100 maf, providing an average annual flow of 10 maf. During this 10 year period, the upper basin must release between 75 and 82.5 maf to the lower basin, leaving at a minimum only 17.5 mar available to the upper basin during this 10-year period. In addition, let us suppose that Lake Powell is full at the start of this drought.

In this event, the total water available to the upper basin in the 10-year period will be 48.5 maf (17.5 maf+31 maf–48.5 maf). This amount would allow the upper-basin use patterns of 1974 to be met; however, it would not allow the planned use patterns of 1995 to be met. These estimates are based on the optimistic view that Lake Powell would be full at the onset of drought. Hcwever, if Lake Powell is not filled to capacity, then less water will be available during an extended drought. For example, in 1974, upper basin storage was only 23.6 maf.

In the event of a 100 year drought that begins with this level of storage, the net water available for use in the upper basin would be only 41 maf (17.5 maf+23.6 maf), or 4.1 maf per year for 10 years. This is even less adequate to meet future needs. Calculations of this sort, carried out using existing records or synthesized records of possible future flows, allow the water managers to estimate the probability that a given water commitment cannot be met in any future year or period of years. If such calculations show that the probability of such a deficit is increasing, then more attention can be given to finding alternative solutions to water-supply problems.

Alternatives for Manipulating Water Supply and Water Use

Even though the Colorado River water-supply system is being placed under stress, all analysts agree that there are alternatives available to cope with that stress. Part of the task of these analysts is to provide the best possible estimates of the probability that supply constraints will be felt. They must also estimate the probable duration of a period of such constraints. Another part of their task is to identify and describe alternatives. Some of the alternatives they have examined include:

1. Changing the existing patterns of land use so that strongly consumptive uses such as agriculture are replaced by less-consumptive uses.
2. Importation of water into the Colorado River system from such sources as the Columbia and Snake rivers in the northwestern United States.
3. Decreasing evapotranspiration; losses by eliminating phreatophytes–water-loving plants of low economic value.
4. Decreasing evaporation losses from canal and reservoir systems.
5. Terminating plans for energy-resource development in the upper basin.
6. Terminating any further withdrawals from the Colorado such as the withdrawals planned for the Central Arizona Project.

These and other alternatives have been presented for analysis by the committees of the National Research Council (National Research Council, 1968, 1977, 1981), and some of them have been examined rather carefully. We cannot consider every alternative, but we can consider some to illustrate the kinds of questions that are raised. The northern states, including northern California, are not yet as close to the limits to be confronted in a drought period as are the states of the southwestern United States, including southern California. Therefore proposals have been made repeatedly to export water from the northern states to the southwestern states. A model for such transfers has been created within the state of California, which has already developed a system that will allow it to transfer water from the north using the California Aqueduct System. Proposals to pipe water across state lines to the Colorado have met firm opposition.

This opposition is recorded in the 1968 legislation authorizing the Central Arizona Project, which contains a provision explicitly forbidding federal studies of diversions from north to south. Water demand in the Colorado River basin could be kept in check by changing the patterns of land use. For example, agriculture in Arizona consumes approximately 90 percent of the water used in the state. Conversion of agricultural land to other uses is already beginning to decrease water use on this land. Such changes can be brought about by changing the fee structure for Colorado River water, which has allowed agricultural users to purchase Colorado River water for much less than metropolitan users. In addition, the cost of ground water use is rising because the water table is dropping. Small farmers cannot afford to pump from deep wells, and they may abandon farming as a result.

Changes might also be brought about through the sale of water rights held by agricultural users. Such a sale might take place if energy-resource development in the upper part of the Colorado River basin promises a greater economic return than agriculture, or if such energy development becomes a critical national need. However, this kind of shift would cause a decrease in water use only if energy development used less water than agriculture. Those who believe that the Colorado River cannot supply all existing commitments during dry periods also argue that the Central Arizona Project (CAP) should be terminated to prevent the planned diversion of 1.2 maf/year from the Colorado River to central Arizona. Analysts have argued that the $2 billion capital costs of this project cannot be justified if water will not be available to feed the system. However, in 1981, construction was begun on this project in spite of these concerns.

Finally, the alternatives of cycling used water back into the system and of augmenting the system with ground water may be considered. Water used for irrigation is partially recycled into the system after use. Unfortunately, this water is usually more saline than the water withdrawn from the river. For example, water taken from the Colorado at Yuma, Arizona, had a salinity of 800 parts per million (ppm) in 1961, but water returned to the river had a salinity of 6000 ppm. Since a 1973 agreement between the United States and Mexico limits the salinity of Colorado River water delivered to Mexico, such recycling faces a severe limit. This has already led to the construction of a desalinization plant near Yuma, which will desalinate water returned to the Colorado from the Wellton-Mohawk agricultural district.

Similarly, augmenting the system by drawing upon ground water has already been found to be a limited solution for two reasons:

1. where groundwater is itself relatively saline, its use only adds to the burden of salt in the river; and
2. the groundwater supply in Arizona is actually being mined-an expression used to indicate that ground water has been taken from supplies beneath Pheonix and Tucson at a rate exceeding any possible rate of replacement of recharge.

As a result, the water table has been dropping for years, creating the specter of unavailability of the very water upon which development of the region was predicated. It was just this decline in groundwater that led to the proposal for the Central Arizona Project. These changes are graphically demonstrated on maps of the entire region that show the decline in water levels and the decreases in the saturated thickness of the aquifer. The history of this decline is shown in a plot of water-level elevation against time. These declines have already had significant consequences.

The cost of raising water from ever-deeper wells has increased steadily, and as a result, there has been a steady decline in the acreage under irrigation. Landowners with water rights to the groundwater resource have been compensated by the federal government for the decline in this resource through a provision in the U.S. tax laws called the depletion allowance. This provision allows water consumers to decrease the taxes they must pay in proportion to the loss of the water resource over which they have control. Since there is no other water resource in the region that can support the high historic levels of agricultural use, the region can be expected to go through substantial changes in land use and productivity.

The Geological Survey has made some preliminary estimates of the rate at which water levels can be expected to decline, given various levels of irrigation use. These estimates, or contingent predictions, can be employed in economic analyses examining the future of agriculture in this region. These predictions have led to discussions of the future possibility of importing surface water from the Great Lakes or the Mississippi River. Thus recognition that the groundwater resource is declining is having the same effect that we saw for the lower Colorado River.

Water Law and Water Policy

In the development of water resources of the Colorado River Basin, government policies and judicial decisions played an extremely important part in determining the allocation of water. Although these decisions were made by the Congress and by federal courts, there is no uniform body of law at the federal level that governs water use and water rights. Most laws governing water use are made at the state level, and this creates a complex situation nationally. Let us look at broad principles and then at some details of law at the state level. Two basic approaches to water law are followed in the United States. One of these is based on the concept of riparian rights; the other is based on the concept of prior appropriation. The concept of riparian rights dominates eastern water law; the concept of prior appropriation dominates much of western water law.

Riparian rights are based on the English common-law view that water is a common-property resource that may be tapped by those who control land along the river bank or lake shore. The same principle applies to groundwater resources; the owner of land overlying an aquifer also owns the right to tap that aquifer and withdraw water from it. The doctrine of prior appropriation takes à different point of view, grounded in Spanish law and practice. Under this doctrine, the right to withdraw and use water is based on historical usage. Someone who puts a certain amount of water to beneficial use establishes the future right to use a similar amount of water. Water is treated as real property, and property rights grow out of prior use or appropriation. Under this system, the initial user is empowered to sell the established water rights to a future user. These two doctrines grew out of patterns of use of surface water resources, but they have been employed in establishing rights to groundwater resources.

Several problems have arisen in many, many regions through the exercise of these rights, so individual states and regions within states have created bodies of law to deal with these problems or have tried to get users to collaborate with each other in managing the resources. We will sample the development of water law and policy in several regions. The most sharply drawn questions concerning the need for water law arise in arid regions where ground water is being mined. Arizona is one such area; Texas and other High Plains states make up another such area.

The dominant water law in these regions is based on the concept of prior appropriation. This doctrine has not limited groundwater use because each land owner has been free to extract as much water as was technically and economically feasible. This practice has created problems for contiguous land owners, since drawing down the potentiometer surface or water table affects a subsurface area much larger than that underlying the property owned by the user responsible for the drawdown. Although this has been known for some time, neither Arizona nor other arid-land states have yet established much water law governing groundwater use. By looking at the development of water law in Arizona, we will see that the original situation is changing. We will also see that water law has a direct effect on the use of land in regions with significant water law. Until 1980, Arizona had no system for controlling the use of ground water.

In June 1980, the Governor of Arizona signed into law a bill controlling water withdrawal and land use. The Governor, Mr. Bruce Babbitt, made the following observation: In the old West, we're going overnight from a laissez-faire system, a system where everybody used whatever they wanted wherever they wanted, to the most comprehensive groundwater management system of any state in the American West. The bill was introduced and passed because the federal Government had informed the state of Arizona that it could not draw from the Colorado River until it had taken action to control its use of water. The new law contained the following provisions:

1. Development is prohibited where the developer cannot assure water availability for the next 100 years, and the introduction of agriculture in problem groundwater areas is also prohibited.
2. Water conservation in major urban and agricultural areas is required.

3. Registration of all water wells is required.
4. A Director of Water Resources is empowered to set per capita consumption limits for cities and to purchase and retire water rights held by farmers effective in 2007.

This kind of regulation was required because prior appropriation and riparian rights had not been adequate to deal with disputes among water users. Other states have also recognized this problem. California has been the site of major controversies and court cases arising from disputes over water rights (Kahrl, 1978). California has dealt with these disputes in two ways. First, it instituted a system of correlative rights, which made the watershed or the aquifer system the basic planning unit within which water use was to be coordinated. Those within the watershed are assigned rights to allocate water derived from the watershed or the aquifer.

At the same time, California developed large-scale water-resource planning, which requires that water be diverted from some regions of the state for use in other regions. This system gives the state the authority to limit the local water use within these regions. Trelease (1977) has observed that water law in the United States has changed and continues to change as new demands are placed on the law and on water resources. Improved scientific understanding continually shows how the withdrawals from surface and groundwater resources have effects spread throughout a larger system, effects that sooner or later must be taken into account in water law and planning. Thus we can expect that there will be a continuing tendency like that signaled in Arizona to develop systems of water-resource control, systems that will demand more and more information for their design.

3

Utilization of Water

In many parts of the world an inadequate supply of water is the most conspicuous and continuing limitation to crop growth. Water relations thus occupy a central place in any survey of developmental physiology. But much of this subject lies outside the scope of a discussion of root function in the soil. Meteorological factors and the characteristics of plant shoots can greatly influence the effects on crops of variations in water supply. These are referred to only briefly in this chapter and attention is directed mainly to the absorption of water by root systems, to the effects of variations in water supply on their growth and function and to their morphological features which contribute to the survival of plants when the supply of water is limiting.

The Driving Force for the Movement of Water

If the water economy of plants is to be seen in perspective, account must be taken of their evolutionary history. They originated in an aquatic environment and photosynthesis, the unique metabolic process on which plant growth depends, requires the absorption of carbon dioxide from the environment with the simultaneous release of oxygen. In submerged plants this exchange occurs in solution but when terrestrial plants evolved, the site of photosynthesis being necessarily in the above-ground parts, the exchange of carbon dioxide and oxygen with an unsaturated atmosphere had the inescapable consequence that water vapour was simultaneously lost. No membrane system was evolved–nor has synthetic chemistry yet produced one which permits the free exchanges of oxygen and carbon dioxide between an aqueous phase in plant cells and an unsaturated atmosphere without the loss of water. Transpiration is thus an inescapable physical consequence of the development of terrestrial vegetation. Studies of the utilization of water by crops entered a new phase with the work of Penman (1948).

Earlier discussions had frequently concerned with the transpiration coefficient; that is to say the quantity of water used per unit of dry matter produced, a physiological relationship between growth and water loss often being implied. Penman, however, showed that, during summer in temperate regions, the quantity of water lost from a complete ground cover of short grass, which was well supplied with water, is about 0.8 of the water loss from the same area of a free water surface; in spring or autumn the corresponding figure was somewhat lower, *c.* 0.7 (Penman, 1948). Somewhat different values, apply to other types of ground cover and Penman's work provided valuable and widely used basis for estimating the water requirements of growing crops. Transpiration by leaves in thus the driving force for the movement of water through plants.

The simplicity of this relationship should not however, obscure the complexity, of processes which can regulate the movement of water within plants, especially when they are subject to a degree of water stress. This matter can be meaningfully discussed only with a thermodynamic background which is here considered only briefly; Slatyer (1967): Kramer (1969); Kozlowski (1968, 1972) and Weatherley (1970) are among the standard sources from which fuller information can be obtained.

Water Potential

The movement of water depends on the existence of gradients of decreasing water potential–that is to say its free energy–which bears no direct relationship to the total quantity of water present. The transfer of water from the soil through plants to the atmosphere is thus appropriately considered in terms of potential gradients. None the less the quantity of water present exerts an important indirect effect; it largely determines how the potential changes when water is introduced into the system or withdrawn. The same basic thermodynamic principles apply to the movement of nutrient ions into plants. However, in the majority of circumstances it is adequate to consider only their concentration in the solution with which roots are in contact, since it is usually sufficiently dilute for the chemical activity of an ion to vary closely with its concentration.

PASSAGE OF WATER THROUGH THE PLANT

Resistances to Flow

The smaller loss of water from a complete vegetation over than from a free water surface, as shown by Penman reflects resistances to the flow of water along its pathway from the soil to the transpiring leaves. Ohm's law provides a convenient analogy which has been widely used since van den Honert (1948) drew attention to its usefulness; if water moves at a constant rate through a system the ratio of the drop in water potential to the resistance to water flow at each step in the transfer process will be constant. Thus, the relative magnitudes of the resistances to water movement along the soil-plant atmosphere pathway, under constant conditions, can be derived from the change in potential. To describe the passage of water realistically the simple Ohm's law analogy must, however, be amplified in two respects. *First,* the major resistances to water flow in the plant can vary in a complex manner. *Second* account must be taken of the capacity of the system, to use another electrical analogy; if increased transpiration leads to a lowering of the water potential in any part of the transpiration stream, water will be withdrawn from the surrounding cells with a corresponding drop in their water potential. It does not, however, indicate all processes which cause water to move in plants.

Although transpiration is dominant under normal circumstances it is not the sole cause of water movement. If plants are maintained in a humid atmosphere the flow of water through them does not usually cease completely; in some species droplets of water may be released by guttation through hydathodes at leaf tips. This process is attributed to gradients in water potential created by osmotic forces; the bleeding sap which can pass out of cut stems or roots, which is sometimes regarded as evidence of the 'active' uptake of water is similarly explained. These processes are not considered further as they account for only a very small fraction of the total water which moves through transpiring plants.

Movement of Water through Plant Shoots to the Atmosphere

When plants are transpiring in a relatively dry atmosphere the decrease in water potential between the interior of the leaf and air exceeds by a considerable factor that in any other part of the system. Between the soil and the transpiring leaf the decrease in water potential is unlikely to exceed about 50 bar in an unwilted plant and is often very considerably less, but the drop is water potential from the interior of the leaf to the atmosphere may approach 1000 bar. It follows, therefore, that the highest resistance to water movement in a plant system can occur in the leaf. The movement of water across the cuticle on the leaf surface is small, except in plants habituated to constantly humid environments and the major gas exchange occurs through the stomata.

The finely balanced mechanisms which control the movement of the guard cells have been the subject of considerable research (see, for example, Meidner and Mansfield, 1968; Monteith and Weatherley, 1976). The two main circumstances which usually lead to stomatal closure are an increase in the concentration of carbon dioxide in the gaseous phase within leaves and a decrease in water potential of leaf cells; these two responses are believed to be linked by a hormonal process in which abscisic, acid is involved. The role of stomata in water conservation is thus evident; when decreasing leaf water potential causes stomatal closure, the loss of water can be much less than that which the prevailing meteorological factors would permit (Lake *et al,* 1969). However; the efficiency of the control mechanism varies widely between species and it reduces water loss to a negligible level only in xerophytes.

The spraying of plants with artificial 'antitranspirants' can restrict water loss, the action of some such substances being to modify stomatal action. The resistance to the upward movement of water through the xylem of the stem and leaf traces is small beside that across the leaf surface. This is well illustrated by the fact that the rate of transpiration is little affected when an appreciable port of the xylem in the stem has been severed by transverse cuts (Mackay and Weatherley, 1973). It is often assumed that the resistance to the passage of water across the leaf varies only because of stomatal movement but there is some evidence that the resistance in the mesophyll may also decrease if the rate of transpiration rises (Boyer, 1974).

Transfer of Water Across Roots

The main resistance in roots to the movement, of water is in its radial passage across the cortex to the xylem. Next to the resistance in leaves it is the principal barrier to water movement throughout the plant; by comparison resistance in the xylem of the root is usually small. This can be easily demonstrated by cutting off roots of plants which are immersed in a solution, transpiration may then increase because water can move directly into the cut ends of the conducting tissue. None the less the resistance in the xylem of the root is not necessarily negligible. Passioura (1972) has suggested that, when grasses are absorbing water mainly from deep in the sub-soil, viscous resistance of water flow in the conducting tissues of their thin roots can be important. Newman (1974) cites evidence which points to the same conclusion of roots exceed *c.* 1 m in length. It is usually considered that water crosses the cortex of roots to the endoderm is mainly, if not entirely, in the free space (Weatherley, 1963; Tanton and Crowdy, 1972), but some workers consider that cytoplasmic movement, or transfer between vacuoles in cortical cells, may make some contribution (Newman, 1974).

When plants of different species are grown in the same environment considerable contrasts in the permeability of roots to water have been detected (Newman, 1973) and there is much evidence that the resistance to the radial transfer of water across a root can vary with environmental factors. Decreased transpiration caused by anaerobiosis, or unfavourable temperature, in the rooting medium, has been attributed to the reduction in the permeability of membranes; the application of respiratory inhibitors can have a similar effect. A review of the effects of temperature by Kramer (1969) indicates that the restriction of absorption caused by low root temperatures is greatest in plants adapted to warm habitats. Variations in the rate of transpiration may also modify the resistance to the radial transfer of water across roots.

Weatherley and his co-workers (Tinklin and Weatherley, 1968: Stoker and Weatherley, 1971) found in several species that if transpiration is increased by a reduction in the humidity of the atmosphere, the resistance to the movement of water across the root decreased. This effect has not, however, been found by all investigators (e.g., Newmann *et al.,* 1974) and some observations suggest that it may occur only at a limited range of transpiration rates (Hailey *et al.*, 1973). Differences in experimental methods may perhaps help to explain discrepancies between investigations. Comparative studies of the entry of water into different parts of root systems provide additional evidence that the radial movement of water across roots is influenced by the rate of transpiration. When the supply of water is ample it enters more rapidly into the apical zones of roots where the endodermis is not suberized (Hayward *et al.,* 1942; Graham *et al.,* 1914). However, experiments by Brouwer (1953) indicate that changes in the rate at which water is absorbed can modify the extent to which it enters different parts of young roots.

The uptake of water by a single root of broad bean *(Vicia faba)* was enhanced either by transferring the plant from dark to light or by placing the other roots in an osmolyte of osmotic potential c.–2.0 bar. Both treatments led to considerably more rapid uptake of water, especially 7.6-12.5 cm *from* the apex; absorption by them then exceeded that in the special 2.5 cm in which the rate of uptake had been greater when transpiration was low. Newman (1974) has reviewed other evidence that the entry of water into different parts of young roots can be varied by external factors but this subject is still very incompletely understood even when the entire root system experiences the same water potential. When plants receive a restricted water supply in soil, further complications may be introduced as a result of discontinuous contact between roots and the soil. None the less, all investigations of the uptake of water by different parts of the root system indicate that water can enter most readily-into the young unsuberized zones of roots, but absorption is not confined to them.

In woody perennial plants, of which over 99 percent of the root system may be suberized, slow absorption through the order tissues is likely to make a major contribution to the total water up-take. Sources or more detailed information on the water relations of plants include Kramer (1969); Hsiao (1973); Newman (1974) and Monteith and Weatherley (1976).

Effects of the External Supply of Water on the Function and Growth of Root Systems

In the preceding chapter the effects of the supply of nutrients on absorption by roots and on the form of root systems was discussed primarily on the basis of observations when roots were grown in solution culture and no other factors were varied. There is no corresponding simple and direct way for studying the effects of differences in the external water potential; it cannot be varied in controlled artifical systems with the same facility as the concentration of nutrients. Evidence on the effects of water potential on root function comes mainly from experiments in soil.

General Relationships between Water Potential and Root Function

The terms permanent wilting point 'and' wilting percentage are used to describe the maximum water potential, or water content, in soil which causes plants to wilt. Slatyer (1967) has pointed out that the permanent wilting point should not be regarded as a soil constant since wilting depends on the loss of turgor from leaves and it is, influenced by the rate of transpiration as well as by the osmotic pressure of the vacuolar sap of cells. Beyond this the practical usefulness of the wilting point concept in field situations is often limited because ample water in part of the rooting zone may permit growth to continue although the soil adjacent to much of the root system is close to the wilting point. It is generally considered that a water potential of 15 bar throughout the rooting zone would lead to the permanent wilting of the majority of crop plants (Kramer, 1969) but the rate of root extension starts to decrease at much higher water potential *e.g.*, often about c–0.5 bar though root extension may continue slowly until the water potential falls to–10 bar or lower (Lawlor, 1973; Newman, 1966).

Some experiments with maize (corn, Zea mays) and tomatoes (Lycopersicon esculentum) indeed indicate that if part of a root system is well supplied with water, some growth can still occur in roots which are surrounded by soil in which the water potential is less than –40 bar, they may remain alive at appreciably lower water potentials (Portas and Taylor, 1976). When osmotic forces contribute appreciably to the water potential, for example in saline soils, or after large addition of fertilizer, the restriction of growth may be greater than if low soil water potential arises from matric forces alone

Osmotic potentials c–2 to –4 bar can be injuriour to many plants, (Kramer, 1969), but halophyles are more tolerant. The injury caused by low osmotic potential may be associated with interference with hormonal processes. Because of the experimental problems of varying the water potential uniformly through the rooting zone, other factors being maintained constant, the effects of water potential on nutrient uptake have been little studied independently of those on growth. However, if low water potential reduces root growth a restriction of nutrient uptake is to be expected because of the reduced metabolic activity of the roots, moreover the metabolic demand of the whole plant for nutrients can be reduced. If osmotic as opposed to matric forces are responsible for low water potential, nutrient uptake may also be affected by competitive processes for example, in saline soils a high concentration of sodium can depress the absorption or other cations.

Effects of Gradients in Water Potential Within the Rooting Zone

Except when the soil is close to or above field capacity appreciable gradients in water potential usually occur. When, water is lost in dry weather its potential usually decreases more rapidly near the surface than in the deeper soil layers. This is due not only to the downward movement of water, but also to its evaporation into the atmosphere and absorption by roots which are often more abundant near the surface. The gradient can be enhanced if gravitational drainage is slower in the deeper layers and sometimes also because their capacity for water retention is higher. Climate, the distribution of roots and soil characteristics all interact in determining the magnitude of gradients of water potential in the soil under growing plants.

Thus, the relationship observed in any set of circumstances cannot be regarded as widely representative. None the less Fig constructed from the work of Garwood and Williams (1967a),

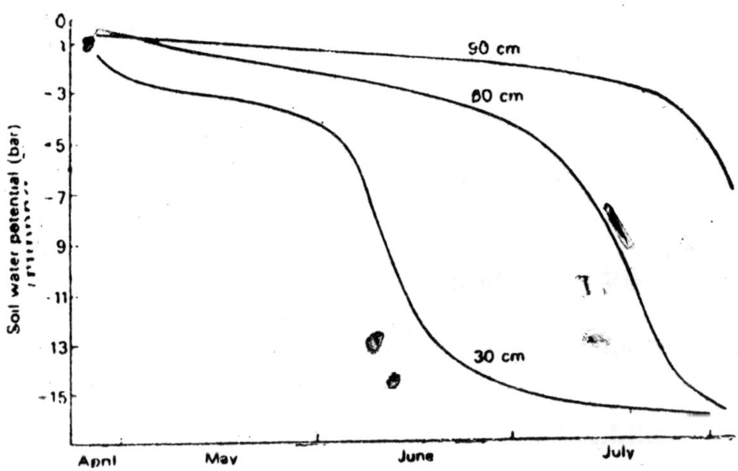

Fig. 3.1. Changes in water potential at three depths in a light alluvial soil underlying an established pasture during three summer months in which the pasture was sheltered from rain. The water potential at 15 cm different little from that at 30 cm during the major part of the period.

illustrates the types of situation which can develop in dry summer weather in a temperate locality; during three summer months a sward of perennial ryegrass (Lolium perenne) in, southern England was protected from rain by a transparent cover at sufficient height to allow free air circulation. A period of drought was thus reasonably simulated. The water potential fell considerably more rapidly at and above 30 cm than at greater depth, and as time passed the uptake of water became increasingly dependent of roots which had penetrated deeply. The general manner in which these conditions affect the distribution of roots and the absorption of nutrients will now be considered.

Root distribution

Under field conditions variations in water supply are frequently the major cause or differences in the distributed of roots, particularly the depth they attain in soil. It illustrates contrasting rooting patterns in comparable soils due to climatic differences while Fig. 3.2 shows variations caused by differences in rainfall on the same field in two successive years. This variability is another example of the capacity of roots of 'compensatory growth' a restriction of growth in part of the root system may lead to increased growth of more favourably placed root members. The analyse adequately the effects of water supply on root growth in the soil, account must be taken not only of the water potential at different depths but also of the rate at which water moves through the roil towards roots; the determines the extent to which water potential is maintained adjacent to the root as absorption proceeds.

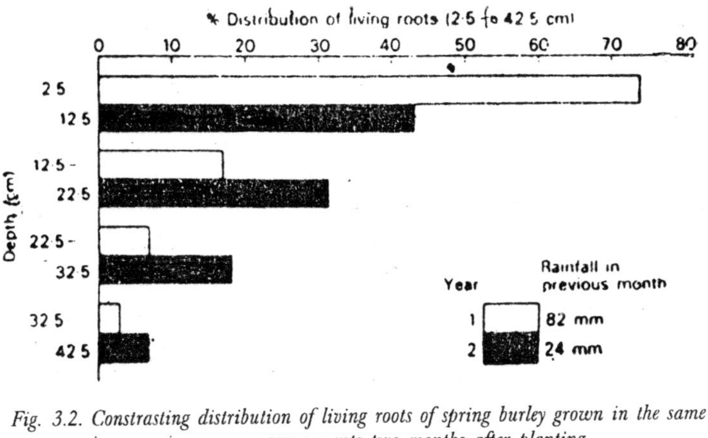

Fig. 3.2. Constrasting distribution of living roots of spring burley grown in the same field in successive years: measurements two months after planting.

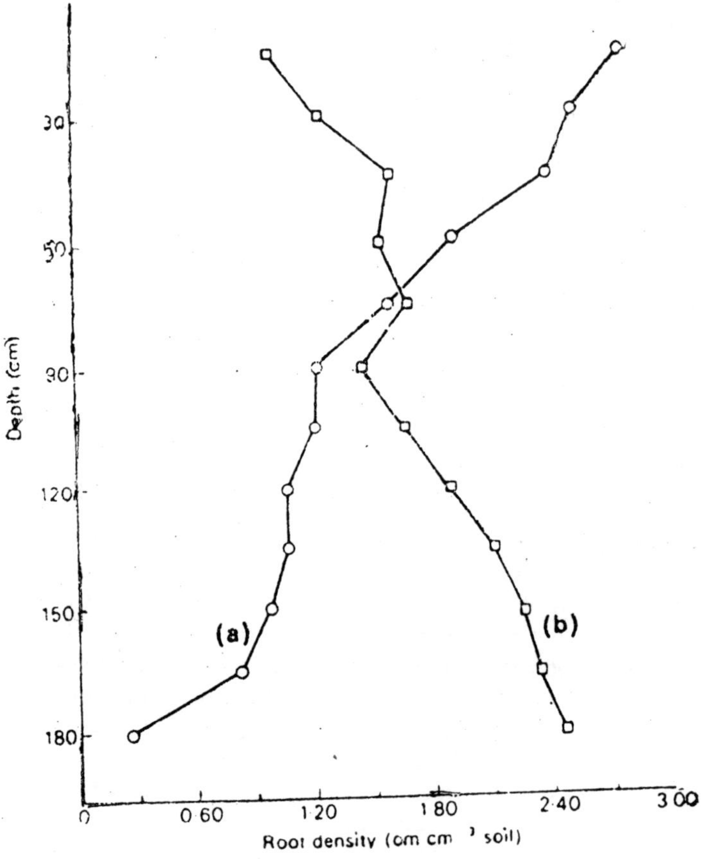

Fig. 3.3. Effect of decreasing water potential for three weeks specially in the appear soil on root distribution in cotton plants (Gossypium hirsutum); (a) at beginning of experiment; (b) at end.

The difficulty of measuring the potential and the quantity of water in soils simultaneously with root distribution has restricted detailed studies of this subject, The Rhizotron at Auburn. Alabama, in which the growth of roots observed in large volumes of soil against glass windows has, however, provided interesting information Fig. 3.3 shows the distribution of the roots of cotton (Gossypium hirsutum) grown in a uniform fine, loamy send of relatively low water retention at the beginning and the end of a twenty-one day period in which no water was added to the soil (Klepper *et al.*, 1973). Initially when the water potential throughout the rooting

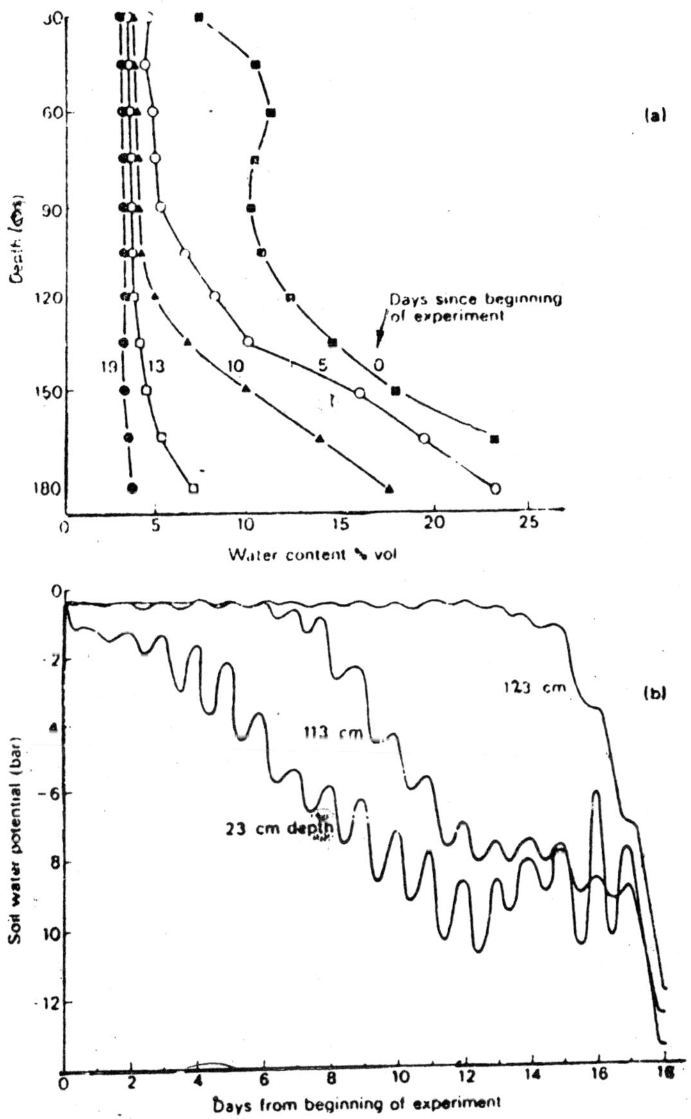

Fig. 3.4. Changes in soil water content (a) and soil water potential (b) at different depths in the soil during the experiment in which root distribution was illustrated in Fig. 3.3. Diurnal variations in soil water potential maximum at 06.00 hours are evident.

depth was c.–0.1 bar the density of roots decreased from the surface downwards. However, during the during cycle when water was lost considerably more rapidly from the surface layers (Fig. 3.4 and b), the pattern of root distribution was reversed (Fig. 3.4).

The density of roots in the dry surface soil decreased due to death by desiccation, while a considerable proliferation occurred in the deeper layers in which the soil water potential fell little until late in the experiment. When plants were grown simultaneously in indentical conditions, except that an ample water supply was maintained throughout the rooting zone, there was little change in distribution of roots during the three week periods. The effects of the desiccation of the upper part of the root system shown in Fig. 3.3 are basically similar to those which can be induced by lowering the solution level when plants are grown in solution culture. The continuing–often enhanced–growth of the lower parts of a root system when water shortage inhibits growth nearer the soil surface has also been demonstrated by Newman (1966); however a response as great as that shown in Fig. 3.3 seldom occurs under field conditions.

Nutrient uptake

Quite short periods of warm, dry weather can cause large variations in the absorption of nutrients from near the soil surface. This is conveniently illustrated by experiments in which soil at different depths has been labelled with radioactive tracers; they enable uptake from different depths to be compared. Fig. 3.5 gives the results of an experiment conducted in this way; variations in the moisture content at the depth of 5 cm in soil under a sward of perennial ryegrass (Lolium perenne) led to corresponding variations in the absorption of calcium from that depth relative to the deeper soil layers, phosphate showed a similar response (Newbould, 1969). In dry conditions the restriction of nutrient uptake from near the soil surface can impose a particular restraint on plant growth because, when the water supply is ample, the surface layers of soil are not infrequently the major source of nutrients for many annual crops as well as perennial grasses; not only is the density of roots greatest in this zone but also nutrients are usually most aundant.

Thus, if the water potential of the surface soil decreases, but ample water remains accessible to the deeper roots, the restriction of plant growth may be largely due to the limited ability of plants to absorb nutrients in this zone; improved plant growth would thus be expected if nutrients were located at a greater depth in the soil even if water supply remained the same. Garwood and Williams (1967b) illustrated this effect (Table 3.1). Swards of Lolium perenne

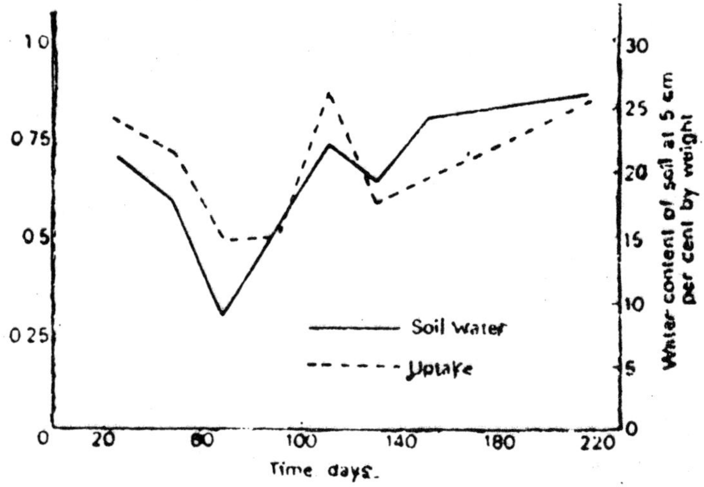

Fig. 3.5. Effect of variations in water content of the surface soil (percentage by weight) during summer on the uptake by perennial ryegrass (Lodium perenne) of calcium from the depth of 5 cm, relative to that from deeper soil layers (10, 20 and 40 cm).

were subjected to drought regimes similar to those used to obtain the water potential profiles; the plants absorbed nitrogen from a fertilizer source considerably more readily when It was placed some depth below the soil surface, whereas when the supply of water was abundant superficial applications were effective. It has sometimes been suggested that if a root experiences a gradient of soil water potential it is able to transfer water from the soil of higher to lower potential. Klute and Peters (1969) have suggested that this does not occur to an important extent.

Table 3.1. Effect of the Depth at which Ammonium Nitrate is Placed in the Soil on the Growth of Perennial Ryegrass (Lolium Perenne) Under drought Conditions.

	Ammonium nitrate equivalent to 112 kg N ha^{-1} applied to all treatments		
Depth of placement of fertilizer (cm beneath soil surface)	0	46	76
Increase in Yield of herbage due to fertilizer: percent	57	82	102
Apparent recovery of fertilizer nitrogen: percent	35	75	80

The Survival of Water Stress

Definition of Water Stress

The term 'water' can convey different meanings depending on whether interest centres mainly on physiological, ecological or agronomic aspects. In the strict physiological sense plants can be regarded as suffering a degree of water stress when cells are not fully turgid–that is to say when their water potential falls below zero. Mild water stress as thus defined occurs whenever plants are transpiring at an appreciable rate; this is implicit in the fact that the flow of water through plants depends on a gradient of decreasing water potential from the roots to the mesophyll of the leaves. However, in environments favourable for plant growth the continued upward movement of water during the hours of darkness, when transpiration has ceased, largely or completely restores the turgor of leaves. Such situations, which are represented diagrammatically are thus normal; none the less, measurements of leaf water potential during daylight hours indicate an appreciable loss of turgor. From the ecological or agronomic viewpoint water stress becomes of concern when it imposes a conspicuous restraint on plant growth.

In dry weather, when evapotranspiration exceeds precipitation and soil water potential becomes progressively lower, the decrease

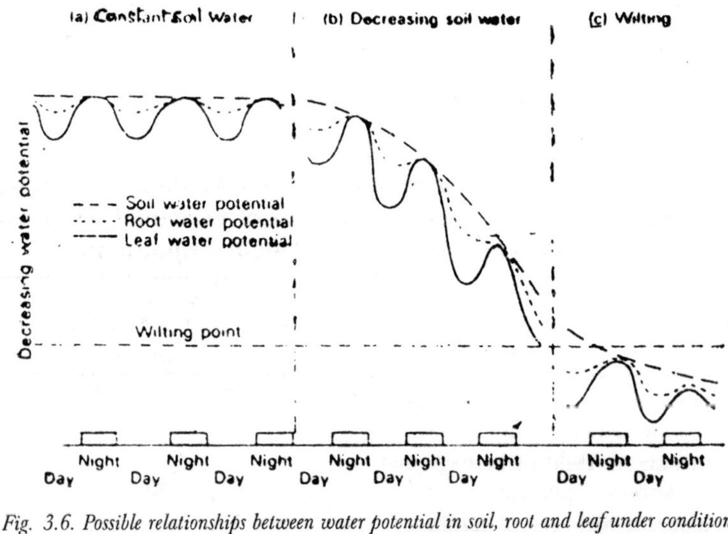

Fig. 3.6. Possible relationships between water potential in soil, root and leaf under conditions of appreciable transpiration when the soil water potential remains high and constant (a), is decreasing (b) and when the wilting point is reached (c).

in the water potential in the plant during the hours of daylight may not be offset by the continuing absorption of water at night. This situation, which is usually accompanied by an appreciable restriction of growth, is illustrated in simplified from in Fig 3.6b. Eventually; if the water potential throughout all or at least the major part of the rooting zone falls to the wilting point growth cases, usually with irreversible injury to foliage. However, regeneration may still be possible.

Characteristics of Shoot which Influence Survival

This discussion is concerned primarily with root growth in relation to the survival of water stress but if this subject is to be seen in 'perspective the role of the above-ground tissue must be borne, in mind, The shoots of no important crop plants possess to a marked degree the characteristics which enable xerophytic plants to withstand considerable periods of desiccation. Note the less, there can be wide differences between species in the extent to which water loss is restricted by stomatal movement in periods of dry weather (Kramer, 1969), In addition, morphological characteristics can greatly influence the ability of plants to recover after relatively severe desiccation.

The regeneration of many pasture grasses after summer drought is, for example, due to the presence of well protected buds near the base of shoots; Even when the water potential in the greater part of the rooting zone has been close to wilting point for many weds in summer, so that growth ceases, a considerable development of foliage can occur in a few weeks after the soil is returned to field capacity. This is well illustrated by the experiments of Garwood and Williams (1967a) with perennial ryegrass (Lolium perenne): growth had ceased before the water potential of the soil was reduced to the low values but rapid growth was resumed within a few weeks of ample water being again provided.

Characteristics of Roots which Influence Survival

The most important characteristic of the root systems of many-crop plants, which contributes to their survival in dry conditions, is the ability of root axes to extend sufficiently rapidly that they maintain continuing contact with zones of soil in which the water potential remains adequate under dry conditions. This is a characteristic which causes lucerne (alfalfa, Medicago sativa) to be among the most drought resistant to crop plants. The importance of root depth in determining water uptake under dry conditions is well from the work of Long and French (1967) who compared water withdrawal from the soil in a period of dry summer weather by meadow fescue (Festuca prantensis), a relatively deep rooted grass, and the more shallow rooted timothy (Phleum pratense). Meadow fescue absorbed considerably more water from below 35 cm, and also a greater total quantity, than the shallow rooted species.

The difference in water extraction between the two species makes it evident that the resistance to the movement of water through the soil to the root zone (sometimes called the pararhizal resistance) can be much greater than the resistance to its movement through the vascular tissue of roots. Next to root depth, extensive root branching is often the most important characteristic of root systems which favours the uptake of water. If there is an ample supply of water throughout the rooting zone the size of root systems may be more than ample to supply the needs of the plant (Newman, 1974) and the removal of an appreciable part of the root system can have little effect on the total water uptake (Andrews and Newman, 1968). But the situation is very different in dry conditions when steep gradients of water potential make it possible for water to be absorbed by only a small part of the root system.

When water stress occurs for only a limited period a further characteristic of the root system can be of considerable importance; namely the rapidity with which new roots develop after the water potential in the soil–especially often the surface layers–again becomes favourable. Many gramineae can rapidly develop new nodal or crown roots in these circumstances and this can contribute to their recovery after periods of drought; but it is not necessarily the only reason. Evidence that the older parts of cereal roots, in which a considerable part of the cortex has collapsed, can absorb nutrients from solutions, makes it seem probable that after a root has been surrounded by dry soil for an extended period the absorption of water, as well as nutrients, can be resumed when the water supply again becomes ample provided that the vascular tissue remains intact. This question does not, however, appear to have been studied detail. To sum up, the combined effect of the depth to which roots penetrate, the extent to which they ramify in zones where water is constantly available and their ability to regenerate when the water supply becomes favourable appear to be the most important morphological characteristics of root systems which enable plants to withstand water stress. But this may not always be so.

Passioura (1974) has described laboratory experiments, with simulated drought conditions, in which plants depend on water stored in the soil throughout their entire period of growth. He found that the yield of grain of wheat (Triticum aestivum) could be enhanced when only a single seminal axis was allowed to penetrate the soil. The benefit of this treatment was attributed to a more economical use of water during the earlier phases of growth which resulted from the higher hydraulic resistance of the restricted root system.

Variability–The Difficulty of Prediction

The response of root systems to variations in water supply under natural conditions is perhaps the most difficult aspect of the behaviour of plants to predict or to describe in quantitative terms. Uncertainty is caused both by the rapid, and largely unpredictable, changes in water supply which seasonal weather can bring about and the equally rapid response of plants these changed conditions–including sometimes injury due to anaerobiosis caused by waterlogging.

Moreover, it is possible that the closeness of contact between roots and the soil, and hence the resistance to the transfer of water, can be affected. The impacticability of even describing the form of the root system of a single species in a single soil in a truly representative manner. Because root density normally decreases with depth it is not surprising that an exponential relationship can often account for a large part of the variation in root distribution with depth in soil. But this is not a constant relationship, it happened in this first (wetter) year but in the second (dry) year the relationship was closely linear. In both years, however, there were considerable discrepancies between the density of root observed at each depth and that calculated from the exponential or linear equations which fitted the observations most closely. Mathematical relationships to describe the growth of plant roots in uniform environments have been developed. However, these models can at present help little in the study of field problems since an understanding of root function is of the greatest interest when they are subject to stresses–often transient and unpredictable ones. The recognition of this in no way disputes the desirability of seeking as fully as possible to derive quantitative relationships which may assist in describing the response of plants to their environment.

4

WATER CYCLE

The most abundant single substance in the biosphere is the familiar but unusual inorganic compound called water. The earth's oceans, ice caps, glaciers, lakes, rivers, soils and atmosphere contain 1.5 billion cubic kilometres of water in one form or another. In nearly all its physical properties water is either unique or at the extreme end of the range of a property. Its extraordinary physical properties, in turn, endow it with a unique chemistry. From these physical and chemical characteristics flows the biological importance of water. It is the purpose of this article to describe some of water's principal qualities and their significance in the biosphere. Water remains a liquid within the temperature range most suited to life processes, yet in due season there are occasions when liquid water exists in equilibrium with its solid and gaseous form, for example as ice on the top of a lake with water vapour in the air above it. Freezing starts at the surface of the water and proceeds downwards; this follows from one of water's many peculiar attributes.

Like everything else, ice included, liquid water contracts when it is cooled, but the shrinkage ceases before solidification, at about four degrees Celsius. From the temperature down to the freezing point the water expands, and because of its decreased density the cooler water floats on top of the warmer. Ice has a density of .92 with respect to the maximum density of water and hence an unconstrained block of ice will float in water with about an eleventh of its volume projecting above the surface. The biological significance of freezing from the surface downward, rather than from the bottom upward, is too well known to need repetition here. Among its other thermal properties water has the greatest specific heat known among liquids (the ability to store heat energy for a given increase in temperature). The same is true of water's latent heat of vaporization: at 20 degrees C. (68 degrees Fahrenheit), 585 calories are required to evaporate one gram of water.

Finally, with the exception of mercury, water has the greatest thermal conductivity of all liquids. Some consequences of water's large latent heat of evaporation, which is a major energizer of the atmosphere, will be considered below. Its great specific heat means that, for a given rate of energy input, the temperature of a given mass of water will rise more slowly than the temperature of any other material. Conversely, as energy is released its temperature will drop more slowly. This slow warming and cooling, together with other important factors, affects yearly, daily and even hourly changes in the temperature of oceans and lakes, which are quite different from the corresponding changes in the temperature of land.

Among other things, this can lead to differences in the thermal regimes of soils that are of major importance in ecology. The type of soil, interacting with water, determines the earliness or lateness of plant growth at a given site; the interaction may also affect the local risk of frost. In basic structure the water molecule has a small dipole moment and is freebly ionized. Water will dissolve almost anything to some extent (fortunately the extent is extremely small for many substances). The dissolved material tends to remain in solution because of another of water's exceptional attributes. The values given by the inverse-square law for the force that attracts separated positive and negative ions are determined by multiplying the square of the distance separating the ions by a constant that varies according to the nature of the separating medium. Known as the dielectric constant, this constant is greater for water than for any other substance. To get the same attractive force in water as in air, for example, the water separation has to be cut down to a ninth of the separation in air. Because of its extreme dielectric constant liquid water in the biosphere is not chemically pure (unlike water vapour, which is always pure, or ice, which can be and often is pure).

Instead liquid water is an ionic solution and one that always contains some hydrogen ions because the water itself can supply them. The concentration of hydrogen ions, expressed as a degree of dilution, gives the physical chemist a numerical index that describes the state of various water samples. The number is the logarithm (to the base 10) of the degree of dilution; the chemist labels pH. For his tests he is armed with a pH meter, calibrated from zero to 14. Fourteen orders of magnitude is an enormous range for any terrestrial property, yet the water content of the soil may give a reading anywhere from pH 3 (very acid) to pH 10 (very alkaline), which is equivalent to a range of from one to 10 million. These are extremes, however, and most terrestrial plant growth-including much of the world's agriculture proceeds in soil with a water content that ranges only a few units on each sides of pH 6.

The range for marine organisms is even more restrictive: coastal waters are about pH 9 and the general oceanic average is just over pH 8. Below pH 7.5 many marine animals die; eggs are particularly vulnerable. Below pH 7 the carbonate in seawater would remain in solution, rendering production of any kind of skeleton impossible. Another method of describing the state of a given water sample is independent of hydrogen-ion content. Material in solution, whether it is ionized or not, disturbs the liquid structure of the water; in thermodynamic terms the presence of solutes decreases the free energy of the water. Many soil and plant workers find it convenient to use the symbol pF for such changes in free energy, with the steps between units also representing one order of magnitude. As with the pH range, the range of pF values is very great. The quantity being measured in pF units is basically a potential, with the same dimensions as pressure. If all the water problems in soils, plants and animals were problems of solutions, it would be sufficient to describe the consequent variations in free energy as variations in osmotic potential, expressed in any of the conventional units of pressure.

The free energy of water, however, can be decreased in other ways, notably in capillary systems. The energy to lift the water into a notably in capillary tube (or in nature into the porous and cellular systems of soils and plants) comes out of the free energy of the water. Today this is known as "matric" potential, a term that has replaced the earlier "capillary" potential. In soils and plants the matric potential may be more than the osmotic potential. A comparison of numbers will give an idea of the pF scale and its ranges. The pressure is expressed as the height in centimetres of an equivalent column of water; thus one bar equals one atmosphere, which equals a 1,000-centimetre water column. This is equivalent to pF 3. In a waterlogged soil, beginning to drain, the matric potential may be between pF 0 and pF 1; in a fully drained soil the potential may be near pF 1.7.

In a soil that is as dry as plant uptake and the transpiration of water from leaves can make it, the matric potential will be about pF 4.2, which is close to 16 atmospheres of suction. The osmotic potential of seawater is near pF 4.5, which makes seawater too "dry" for plant roots: the salt content of plant cells might be anywhere in a range from less than pF 4 to pF 4.5. Here once again water is extreme. Associated with the matric potential in a capillary system there is a curved liquid-air interface; the value of the potential is found by doubling the known value of the liquid's surface tension and dividing the product by the radius of curvature.

Water has the greatest surface tension of any liquid known, so that at any given matric potential the radius of curvature of a water meniscus will be greater than it could be for another liquid. The greater the radius of curvature, the greater the total water content. In a soil this means that more liquid can be retained as water just because it is water. In general, but not always, this is an advantage for plant growth. The effects of a decrease in the free energy of water contained in porous soils or in the tissues of plants include a lowering of the fluid's freezing point and vapour pressure. If the source of the decrease is a matric potential, there is also negative pressure, or suction, that tends to pull all kinds of retaining walls together. The effect of freezing in soils and rocks is worth a brief aside.

As the temperature falls the water in the larger soil pores freezes first and the free-energy gradient is such that water will be withdrawn from the smaller pores. As a result ice lenses form in the coarser pore spaces and the finer pore spaces are exposed to greater shrinkage forces. Because water expands on freezing, the ice lenses have a disruptive effect as they make room for themselves. In rock this is the beginning of one method of soil formation. There tends to be a preferred size for the rock fragments produced by ice disruption. This size is near the optimum for transport by wind and is the dominant size in many of the loses soils that have accumulated in areas near glaciers.

In soil the ice disruption is the source of "frost tilth," which is sought by farmers when they leave land roughly plowed in the fall and hope for a sufficiently frosty winter. There are still some uncertainties with respect to the world's water balance, but agreement was reached on probable values or ranges during an international symposium on the subject held in Britain this summer as one of the activities of the International Hydrological Decade. The figures that follow are taken from the proceedings of the symposium. The world's water exists as liquid (salt and fresh), as solid (fresh) and as vapour (fresh).

There is some uncertainty in the value of the total volume, but it is near 1,500 million cubic kilometres. Estimates of the components are most easily expressed as average depths per unit area of the entire surface of the earth, which has a total area of 510 million square kilometres. Oceans and seas-liquid salt water-make up about 97 percent of all water, with an equivalent depth of between 2,700 and 2,800 metres; the greater part is in the Southern Hemisphere. Of the remaining 3 percent, three-quarters is locked up as solid in the polar ice caps and in glaciers. Here measurement is quite difficult, and a spread in estimates is inevitable. The equivalent depth of ice and snow may be near 120 metres, but at the recent symposium a value of 50 metres was not challenged.

The other large component of fresh liquid water is subject to similar uncertainty: the estimates for underground water may be near 45 metres, but again a value near 15 metres was not challenged. Estimates for surface water, mainly in the great lakes of the world, ranged from .4 metre to one metre. There is general agreement on the average water-vapour content of the atmosphere, at an equivalent in liquid of .03 metre. Although this is a very small fraction of the total, size is no measure of importance. Without water in the atmosphere there would be no weather: Leonardo da Vinci's dictum, "Water is the driver of nature," is justified on meteorological grounds alone. A little detail at this point will be helpful as an introduction to another aspect of the world circulation of water. The amount of water vapour is not the same everywhere, either geographically or seasonally. It is greatest at and near the Equator. If the air there were squeezed dry, it would yield about 44 millimetres of rainfall.

In middle latitudes, say from 40 to 50 degrees, the summer yield would be near 20 millimetres and the winter yield near 10 millimetres, with large variations that depend on geography and weather patterns. In the polar regions the yield ranges from two millimetres in winter to as much as eight in summer. Water vapour enters the atmosphere by evaporation (this term includes transpiration by vegetation), and the main oceanic sources are fairly identifiable. It leaves the atmosphere as rain or snow, and because the precipitation may take place close to the source or thousands of miles away, the residence time may vary from a few hours to a few weeks. A general average is nine or 10 days. The general balance of evaporation and precipitation needs three sets of figures, one set for the entire earth, one for the oceans and one for the land surface.

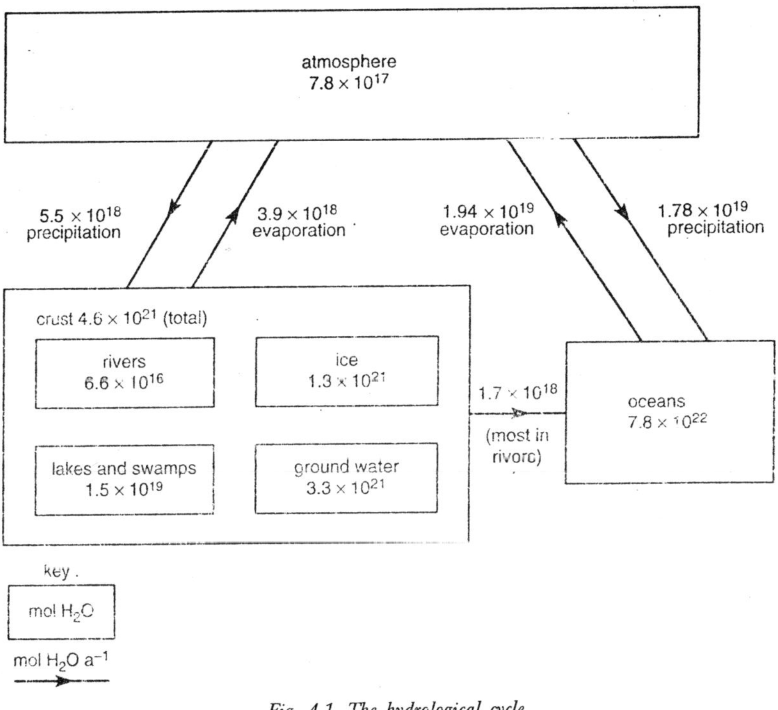

Fig. 4.1. The hydrological cycle.

Here, within a few percent, there is almost complete agreement on values. For the entire earth, average evaporation and precipitation are equal–as they must be–at very nearly 100 centimetres per year. For the oceans, expressed as equivalent depths over the area of the oceans, the average annual precipitation is between 107 and 114 centimetres, the average annual evaporation is between 116 and 124 centimetres and balance is restored by river flow, with an annual very close to 10 centimetres in all estimates. For the land surface the average annual precipitation is near 71 centimetres, the average annual evaporation is near 47 centimetres and the average annual river discharge is near 24 centimetres. (The ocean figure of 10 centimetres corresponds to the 24-centimetre land figure.) Because half of the land surface–ice caps, deserts, mountains, tundra–contributes little or nothing to evaporation, a better evaporation average would take into consideration only the land component of the biosphere where the availability of water is combined with the opportunity for evaporation. Here the average evaporation may total 100 centimetres per year.

The evaporation in high latitudes would of course be far less that the evaporation nearer the Equator. Available measurements support this conclusion. In Finland, at 65 degrees north latitude, the average evaporation is 20 centimetres per year; in southeastern England, at 50 degrees north, it is 50; in North Carolina, at 35 degrees north, it ranges from 80 to 120. On the Equator in the Congo basin the average is 120 centimetres per year; at the same latitude in Kenya it is 150. In the papyrus swamps of the Nile in the southern Sudan, 10 degrees north of the Equator, the average is 240 centimetres per year, but this is a special case. Here the river carries its water into the desert environment of the Sudd; evaporation rates are high not only because of the clear skies and intense sunshine overhead but also because the surrounding desert is a source of hot dry air that augments evaporation. This kind of advective augmentation operates in many places other than the Sudan, particularly in semiarid regions where irrigation is practiced, and not quite enough is known about it. Once in the air, water vapour may circulate locally or become part of the general circulation of the atmosphere.

The general circulation is one of the three important ways of moving water across the earth. Some indication of the worldwide volumes involved is given by the fact that the total annual precipitation over the U.S. comes to some 6,000 cubic kilometres, whereas the liquid equivalent of the water vapour that passes over the U.S. in a year owing to the general circulation of the atmosphere is 10 times that amount. Of the two remaining important ways of moving water across the earth, the major ocean currents comprise one and the discharge of rivers comprises the other. Both have substantial effects on the biosphere.

The ocean currents carry energy surpluses or deficits over great distances; one well-known instance accounts for the extreme contrast between the climates on the west and east sides of the Atlantic in the areas between 50 and 55 degrees north latitude. Without the Gulf Stream northwestern Europe would be a much less pleasant place in which to live and work; indeed, if the cold Lobrador Current had replaced the Gulf Stream, the history of civilization would have been very different. The rivers of the world not only are long-distance movers of water but also serve as conduits for dissolved and suspended material.

Because of its chemical and physical properties, water is a very efficient erosive agent; erosion, transport and deposition have to be recognized as geological processes associated with water in the biosphere. They are the processes that have produced lands and soils, now densely populated and intensively cropped, where annual floods and silt deposition are regarded as the mainstay of life. Elsewhere, notably in the Americas, silt is an embarrassment in the deltas where it settles, and its production is equally unwelcome in river headwaters. Two further points about river waterdeserve mention. First, the salt content of river water differs markedly in composition from that of the oceans. This suggests that the oceanic brine is not merely the accumulation of salts from aeons of land-surface leaching. Second, information about river discharge rates is scanty and not always reliable. As an example, it is only recently that a good estimate of the flow of the Amazon has been obtained. It proved to be twice the best previous estimate and indicates that almost a fifth of the world's river discharge comes from this one stream. It is not possible to do more than guess at the average amount of water the world's plant and animal populations contain.

Considered as the equivalent of rainfall, it may amount to about one millimetre over the entire surface of the earth. This is less by one order of magnitude than the amount of water vapour in the atmosphere, and its distribution is even more varied in space and time. For a fully grown good crop of corn in North America or of sugar beet in northwestern Europe the amount might come to the equivalent of five millimetres of rainfall, and its summer residence time would be two or three days. This is a measure of the rate of water supply needed to maintain optimum conditions for growth. Here, at the point of water untake by the roots of plants, begins the problem with respect to water in the biosphere that makes all other water problems seen trifling. With unimportant exceptions, the basis of all life on the earth is photosynthesis by green plants, a process that involves physics (in the fixation of solar energy) and chemistry (in the union of carbon dioxide and water to form carbohydrates and more complex biochemical compounds).

Water comes into the story in two ways: in transit (as part of the transpiration stream) and in residence (as its hydrogen is chemically bound into the plant structure). The amount that is bound, however, may be less than a fifth of the amount in transit. To give scale the argument that follows, here are some values based on a real crop in a real climate. In producing 20 fresh-weight tons of crop, 2,000 tons of water will pass into the plants at their roots. At harvest perhaps 15 tons of the water supply will be in transit, leaving the crop with a dry weight of five tons. To produce the five tons of dry matter three tons of water will have been fixed and transformed. The energy fixed in the dry matter will be 1 percent or less of the total solar energy received by the crop; nearly 40 percent of the energy will have been used to evaporate the transpired water. The average value of 40 percent for the net solar radiation income retained by a green crop cover varies, of course, with season and climate. The first loss is to reflectivity: of the solar radiation reaching the crop about 30 percent is reflected. There is also an income of long-wave radiation from the sky, but this is outweighed by the outgo of long-wave radiation from the earth to the atmosphere. When the deficit is met by deducting it from the remaining balance of short-wave solar income, the net retained income is decreased to 40 percent of the initial input.

As already noted, when the water is available, very nearly all this energy is used in evaporating water. Here once again water stands at the extreme of a range of physical properties. The volume of water evaporated per unit of energy input is less than it would be for any other liquid. The relevant physical constant, the latent heat of vaporization, is somewhat less than 600 calories per gram at ordinary temperatures, but the rounded figure is adequate for the present purpose. If we let R_1 represent the total radiant income in calories per square centimetre over a period of time, then the net radiation is about $4R_1$ and the evaporation equivalent is near R_1/- 1,500 grams per square centimetre (or centimetres of water depth as the equivalent of rainfall). Consider some real midsummer values to show what this means.

In a humid temperate climate the value of R_1 is close to 450 calories per square centimetre per day. This works out to an evaporation equivalent of three millimetres per day, which is a good estimate for June in southwestern England. For many of the farming areas of the U.S. The R_1 value is close to 650 calories per square centimetre, bringing the evaporation rate up to about 4.5 millimetres per day. The maximum rates known, which are found in irrigated areas, range from 4.5 to 7.5 millimetres per day. It is possible that the higher rates are influenced by advection from surrounding nonirrigated areas, as is the case in the papyrus swamps of the Nile. The most important fact to be considered in connection with this wide range of evaporation rates is that there are only very small variations among the evaporation rates of different kinds of plants. Thus the governing factor in variation is almost exclusively a climatic one. This fact and much other evidence suggest that the supposed water "need" of a crop is dictated not by the plants but by the weather. In this connection the concept of "potential transpiration," which came into use simultaneously and independently in at least two parts of the world, is of great value both in research and in the practical aspects of soil water management.

It is worthwhile seeing how potential transpiration is linked with elementary plant physiology and with some of the physics of soil water already considered. A growing plant takes in water at the roots and, in the absence of immediate replenishment, the process dries the soil so that more and more energy is required for further extraction. The energy requirement is very small, however, compared with the amount of energy needed to evaporate the same quantity of water from the plant's leaves. There can be no serious error in assuming, as Frank J. Veihmeyer of the University of California at Davis does, that all soil water is equally available for transpiration up to the stage marked by the onset of wilting.

The purpose of well-managed irrigation, of course, is to make sure that plants never get to the wilting stage. For maximum growth irrigation may have to consist of frequent small applications of water rather than occasional large ones. Given an adequate supply of water, the chain of consequences is simple. There are maximum values for each of several factors: water content in the plant, hydrostatic pressure in the plant and leaf turgidity. When neither the intensity of the light nor the concentration of carbon dioxide constitutes a limiting factor, maximum leaf turgidity permits maximum opening of the stomatal apertures in the leaf surface, thus affording the best possible opportunity for movement of carbon dioxide into the leaf. The state of the stomatal opening that allows easy inflow of carbon dioxide, however, also allows equally easy outflow of water vapour.

By far the greater part of the water need of plants is actually a "leakage" process that has to be kept goring to ensure continued growth. Given a sufficiently wet soil around the plants' roots, the rate of leakage is dictated not by plant physiology but by the physical factors of temperature, humidity and ventilation. The sole constraint is imposed by the law of energy conservation. In its last stages the transpiration stream undergoes a change of state from liquid to vapour, and the rate of change depends on the rate at which energy can reach the system to supply the necessary latent heat of vaporization. So much for the physics of the process. When the supply of water in the soil approaches exhaustion, plant physiology rather than physics begins to predominate. Plant type, root structure, phase of plant development, soil type, soil depth-these become the important factors. What is available for utilization has more significance than the weather has, particularly in semiarid zones. Because agriculture is most active in the more humid zones of the biosphere, it is useful to estimate how much reserve soil water is available on the average in these zones. Factors already described prevent any exact answer to this question.

Nonetheless, a cautions estimate, advanced with considerable reservation, would be about 10 centimetres of rainfall equivalent. Three examples will suffice to show the need for caution. There are large agricultural areas of North Carolina and neighbouring states where an inert subsoil is covered by no more than 20 centimetres of useful topsoil. Here the entire water reserve available to the agricultural cycle cannot exceed a rainfall equivalent of five centimetres. This is one extreme, the deep volcanic soils of East Africa are at the other. In those soils the roots of many plants go down as much as six metres below the surface. The available water in a profile that deep is equivalent to nearly 50 centimetres of rain, and the plant can extract water throughout a long dry season at something very close to the potential transpiration rate. An example from France falls somewhere in between. There the drying of the soil was observed while a crop of sugar beet transpired at the full potential rate throughout a dry summer.

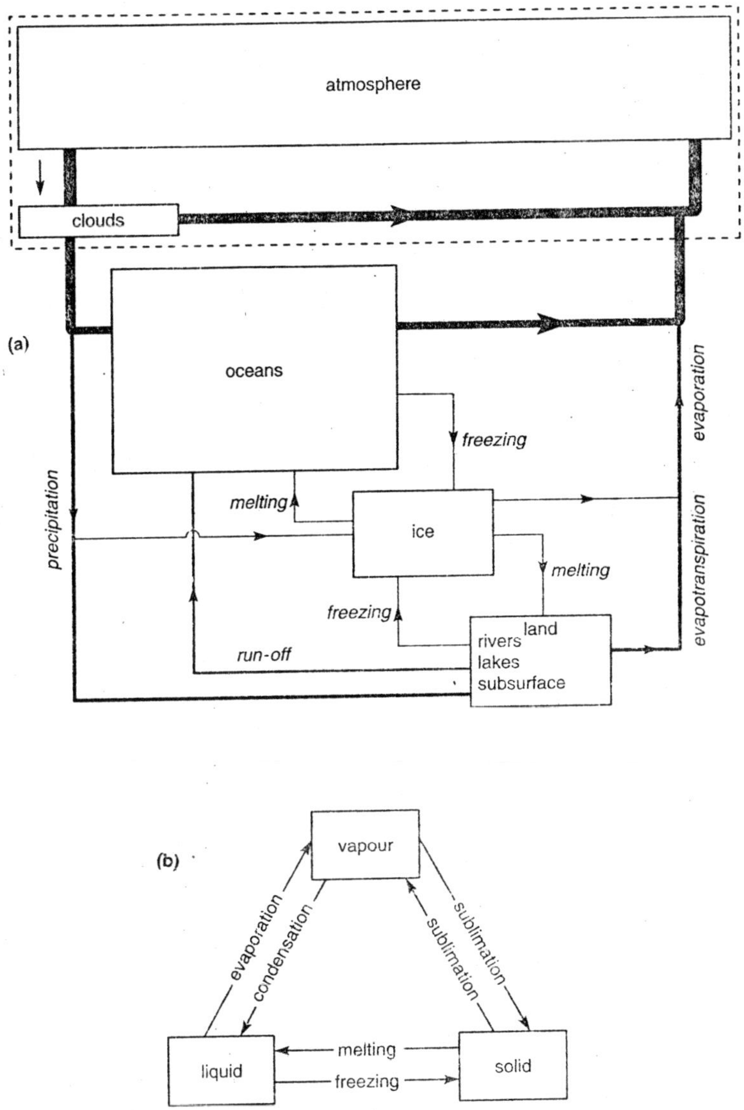

Fig. 4.2. Processes in the hydrological cycle: (a) the hydrological system; (b) changes in phase occurring in the hydrological cycle.

At the driest stage the crop had withdrawn from the soil available water equivalent to 27 centimetres of rainfall. Soil water and ground water are closely related, but whereas soil water is always biologically important, the importance of ground water may range from being trivial to being all that matters. The soil is a kind of buffer between rainfall and ground water. In general any deficit in soil moisture that has built up in a dry period must be completely restored by rain before there is any water surplus available to move down to ground water. This is an important consideration for the water engineer, who may be drawing a water supply from a stream (permanent streams are sustained by ground water) or may be tapping an aquifer directly by means of a well. In the first instance the engineer will presumably have to work within legal constraints on how much river water he can divert.

In the second, if he is to choose a safe aquifer pumping rate, the engineer must (or should!) have some awareness of the current soil-moisture deficit and of the likely rates of rainfall in the months ahead. The river engineer can use the same information for another purpose: the soil-moisture deficit will enable him to estimate the risk of flooding in the event of a heavy storm. In some countries the control of ground water is a major outlet for engineering skill, and its exploitation is the basis of farming technique. One need only think on the one hand of the Netherlands and on the other of such semiarid regions as Iran, where deep tunnels tap the buried aquifers and carry ground water to valley bottoms.

In many semiarid regions the vegetation along transient streams maintains its luxuriance because the groundwater level there is close to the surface and within reach of plant roots. The plants' effective reach depends both on the soil and on the kind of plant, but in general it is seldom more than a few metres. The movements of any water table deeper than that is unaffected by the plant growth or the evaporation processes taking place above it, and its ground water contributes nothing to the biological activity at the surface. What has been said about water so far has involved terms that are generally accepted, and the concepts themselves are supported by good reasoning, good evidence or both.

The remarks that follow, although also based on reasoning and evidence, are more speculative and personal. If the biosphere is taken to be the place where water and energy interact, can the interaction be expressed quantitatively in terms of biological productivity? The answer has to be no. There are too many variable. All the same, by rearranging some of the water quantities and energy quantities that are known, a suggestive relation can be obtained. Start with the fact that, for a good crop, 1 percent or less of the incoming solar radiation is fixed as dry matter (here and in what follows that 1 percent refers to the total botanical yield, irrespective of economic value). We shall give this percentage the symbol e, and express it numerically as 100 per 10,000. This degree of efficiency is achieved only by an experiment station or by an extremely competent commercial farmer.

Based on the statistics of world cereal production, including straw as well as grain, the average achievement in highly mechanized industrial farming shows an efficiency of only about 35 per 10,000. This decreases to roughly 17 per 10,000 in North America, and in tropical Africa and Asia subsistence farming rarely shows an efficiency better than 8, even when the possibility of two crops per year is allowed for. There is obviously room for improvement everywhere and the question in the present context is: Where does water come in, and how? Some evidence is now being accumulated suggesting that, when there are no limitations on water supply, the total crop growth is proportional to the total of potential transpiration over the period of growth. The factor of proportionality depends on many things: plant variety, management, kinds and quantities of fertilizer, pest and disease controls and the like. Hence the negative answer to the earlier question. Still, something can be done with ratios.

There is reason to believe potential transpiration is a fairly constant fraction of the solar radiation income. Combining this fact with the relation between potential transpiration and total crop growth, it is possible to derive a connection between growth rate and utilization of water, assuming that unlimited water is available (by inference, there would be a similar response to timely irrigation). In what may seem too precise a form, the answer is that the increase in yield (t) equals .39 ε. Here ε represents efficiency and t can be read, according to preference, as metric tons per hectare per centimetre of water applied or as tons per acre per inch of water. Considering the fivefold (or perhaps tenfold) world variation in efficiency, some uncertainty in the multiplying factor is unimportant; others may prefer, or find, a different value. Taking the illustration already given, suppose the area involved is one acre and the efficiency is exactly or almost 100 per 10,000.

The predicted increase in yield in response to water, applied when it is needed, is .39 ton of dry matter per acre per inch of water. In terms of fresh weight the gain is about 1.5 tons per acre per inch. This is the kind of response obtained in experiments with irrigated potatoes in Britain. There are some countries where the value of ε is small because of lack of water, but there are many, including several of the rice-growing nations, where the small value of ε is more truly a measure of the inefficiency of the farming system itself. To get the most out of water, whether it comes from irrigation or from rainfall, the standard of performance elsewhere in the system must be improved: better varieties, better soil management, better crop husbandry, better plant hygiene and better pest control. Then water may be the driver of nature in agriculture as well as in the atmosphere.

5

HUMAN IMPACT ON THE SEA

The oceans are truly the last frontier for human exploration and exploitation on this planet. For years, the vastness and volume of the oceans; coupled with the fact that humans are terrestrial organisms ill-equipped naturally to enter into water more than superficially, kept the oceans and their communities safe from significant human interference or effects. As the human population of the world increased through the centuries, large land areas were significantly altered, but the oceans remains relatively untouched. The major usage of the sea during these centuries was as a source of food but the primitive nature of gear and the limited range of vessels combined to make the effects of fishing insignificant. In the twentieth century, all this has changed. The explosion of technology has enabled human beings to permeate all puts of the oceans. The rapidly increasing human population coupled with sophisticated technology, have in a few short decades had significant effects on the ecology of the world's oceans. The purpose of this chapter is to discuss and evaluate some of these effects in light of our knowledge of the functioning of the oceans' ecosystems.

FISHERIES

The earliest use of the oceans by humans was likely for food. Early human populations living along the ocean captured various shore fishes and selfish for consumption. This is recorded in the various large "shell mounds" excavated by archaeologist in different parts of the world. With the advent of vessels to venture onto the sea and the development of more refined nets fishing became an important source of food to people living near the oceans. However, the gear and the vessels were sufficiently inefficient through much of human history. So that even as late in the 1880s, scientists such as T.H. Huxley believed that the major fisheries were inexhaustible.

In the decade of the century, however, the old shins and gears have been replaced by much larger and more powerful vessels, more effective nets and traps, and electronic devices for detecting fish schools. The result has been a significant reduction in many fish populations and the disappearance or overexploitation of others at a time when increasing human populations are demanding more food. Since many believe the oceans are the major food reservoir for upcoming generations of human, the impact of fishing and its potential to sustain future generations bear consideration here.

Major Fishing Areas

Geographically, the major fisheries are concentrated in the waters overlying the continental shelves around the world. The only major fisheries that operate in the open oceanic regions of the world are those for tuna and whales. There are several reasons for the concentration of fisheries in neritic waters. In the first place, the inshore waters have a much higher primary productivity than most open ocean waters and therefore support larger populations of fish's at all tropic levels. Secondly, the bottom is fairly shallow on the shelf and is accessible to the various nets and traps used by humans to capture fishes.

The deep ocean floor, on the other hand is so far removed from the surface that even with our current sophisticated capture mechanisms, we cannot commercially fish there finally. The great lack of food in abyssal depths precludes any large fish population from existing there. Any fishes there could likely not sustain a fishery for any length of time. Although the fisheries are concentrated in the shelf regions, there is a very unequal distribution of the tonnage caught in various areas. The shelf areas of northwest Europe, along the upwelling coast of western South America, and off Japan produce the largest catches of fishes. The southern oceans and the tropics with the exception of the west coast of South America contribute much less to world fisheries.

Major Commercial Species

The commercially important marine animals are drawn from four groups bony and cartilaginous fishes, marine mammals, molluscs, and crustaceans. Of these groups, the various fishes constitute by far the greatest tonnage. Among the thousands of species of marine fishes only a very few make up the majority of fisheries catches throughout the world. These can be assembled into a few major groups (Table 5.1). The herrings sardines, and anchovies account for the largest tonnage of fishes of these a single species, the Peruvian anchoveta (*Engraulis ringens*) accounts for almost half the catch and is the basis of the largest single fishery in the world off Peru. Significant other fisheries for this group are for menhaden off the Atlantic and Gulf coast of the United States and the herring fishery of northwestern Europe. Before its collapse in the 1940s and 1950s, the California sardine fishery was an important contributor to tonnage in this group.

Herrings, anchovies and anchovetas, commonly called clupeiod fishes are small pelagic and feed on a low tropic level either directly on phytoplankton (Engraulis) or on herbivores such as copepods. Most are not used as food for humans, but are reduced to protein meal for animal food. The second largest group of fishes landed are the gadoids, comprising various cod, haddock, Pollock and bake. These are bottom-dwelling (demersal) fishes and the major fisheries are on the shallow banks of the North Atlantic and North Pacific oceans. The third largest groups are the scombroid fishes, generally known by the common name "mackerel". Fishes of this group are common in both temperate and tropical water, where they are fast-moving pelagic carnivores in shallow water. Closely related to mackerels are the various tunas.

Tunas are among the largest fishes in commercial fisheries. They are also the basis of the only major open-ocean fishery. They are widespread, fast-swimming carnivores of tropical and warm temperate seas. Although the tonnage is not as great as for other fishes, they are significant because they command a high price as a canned fish product. Redfishes, rockfishes, and sea basses are demersal, cold-water fishes which are primarily used as human food. The flatfishes such as halibut sole plaice, and flounder, are all well-known fishes that are of considerable economic importance because they are premium eating fishes. All are caught in fairly shallow waters and many have been heavily exploited for years.

The sharks and other cartilaginous fishes are generally not consumed directly due to the high urea content of the flesh. One group not separated in Table 5.1 is salmonids. They form a very minor part of the world landings, too small to merit separate listing, but are of great economic importance in North America and Europe because of their fine taste.

Table 5.1. The catches of different groups of fishes from 1971 to 1977 (numbers are in millions of metric tons)

	1971	1972	1973	1974	1975	1976	1977
Flounders, halibut, sole	1.39	1.29	1.24	1.18	1.14	1.12	1.08
Cod, hake, haddock	10.66	11.44	11.96	12.68	11.85	12.14	10.69
Redfish, basses, congers	4.07	4.16	4.26	4.79	4.97	4.86	5.14
Jacks, mullets, sauries	4.06	4.95	5.64	5.35	5.85	7.25	8.68
Herrings, sardines, anchovies	19.81	13.36	11.41	14.05	13.75	15.03	12.96
Tunas, bonitos, billfish	1.81	1.96	2.09	2.25	2.09	2.29	2.33
Mackerels, snocks, cutlass fishes	3.27	3.18	3.43	3.61	3.61	3.30	3.56
Sharks, rays, chimaeras etc.	0.52	0.53	0.60	0.56	0.59	0.56	0.59
Misc marine fish	8.19	8.65	9.13	8.81	8.55	8.92	8.92
Shrimps, prawns etc	1.04	1.11	1.26	1.32	1.29	1.41	1.44
Krill, platonic crustacean	0	0	0	0	0	0.005	0.0001
Crabs	0.36	0.37	0.39	0.41	0.38	0.40	0.45
Lobsters	0.095	0.094	0.096	0.093	0.095	0.099	0.096
Oysters	0.73	0.78	0.79	0.74	0.84	0.90	0.88
Mussels	0.64	0.52	0.44	0.38	0.46	0.48	0.54
Squid, octopus etc	0.91	1.15	1.07	1.07	1.17	1.17	1.16
Clams, cockles,arkshells	0.66	0.68	0.68	0.73	0.70	0.76	0.83
Scallops, pectens	0.16	0.89	0.02	0.24	0.28	0.36	0.41

The major crustacean group in world fisheries is shrimp, various species being caught in both warm and cool waters in all oceans. The other significant crustaceas are various species of crabs and lobsters. All are economically valuable and command high prices for direct human consumption. The major mollusc group in terms of tonnage is the squid. Significant numbers of squids are caught in Japan, Europe, and California. The other major mollusk group

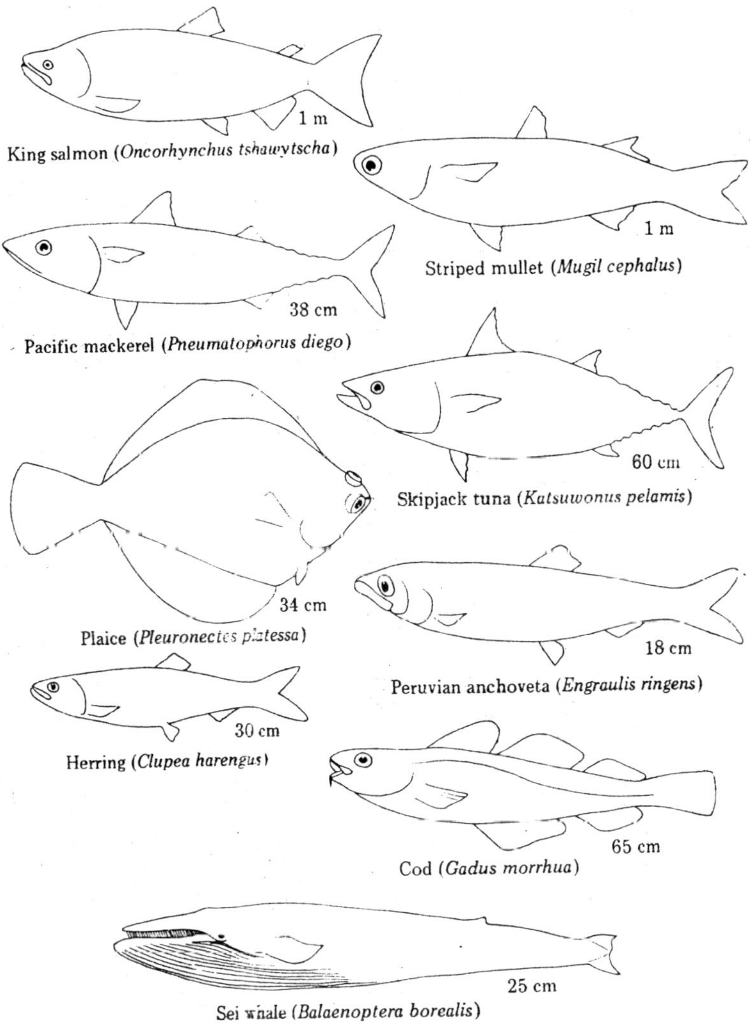

Fig. 5.1. Examples of some of the major commercial fishes of the world.

commercially harvested includes the bivalves, such as oysters and clams oysters and clams. The major fishery for marine mammals has been that for various species of whales. Persistent over exploitation of the stocks in this century has. However, Virtually doomed this fishery. Other harvestable mammals are mainly seals and sea lions. Other minor fisheries exist for various species of algae. Some are taken in Japan for human consumption, but most are collected to extract various products for use in industry (algin derivatives).

Sustainable Yield and the Future

In 1977, the total world fishery catch was 73.5 million metric tons, of which 50.9 million metric tons were for human consumption. Of the 73.5 million metric tons, 62.7 million metric tons were from marine fisheries. The total fishes caught for human consumption represents perhaps 1 percent of all human food. But a significant 10 percent of the protein intake. Hence its importance, the total marine tonnage in 1977 was down from 1976. Which landings were a record 64.1 million metric tons. In fact the recent landings have all been relatively stable compared to the rapid growth in the 1950s and 1960s, since the human population is still increasing and is expected to reach near 7 billion by the end of the century.

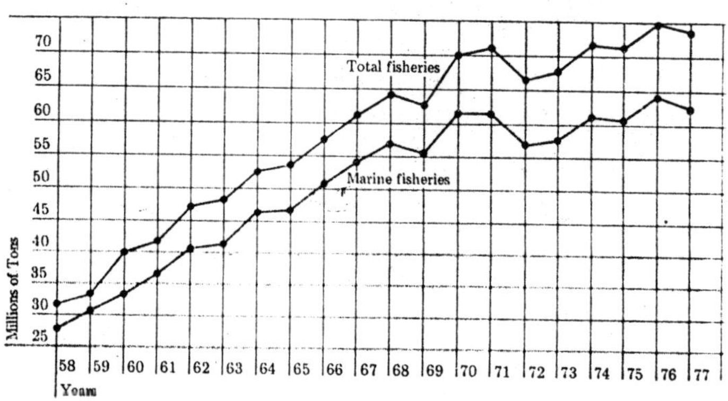

Fig. 5.2. Changes in the world annual fish catch in recent years.

And since many are anticipating that much of the food needed to, feed these increased numbers will come from the sea, does this leveling off mean we have reached the limits? What are the facts, with respect to current and potential amounts of food from the sea? In order to answer these questions. It is first necessary to look briefly at the current fisheries situation, of the present fishing effort. 90 percent is concentrated on the 7 percent of the world ocean area represented by the continental shelves; these shallow areas are the most productive ocean areas.

The great growth of the world fisheries in the 1950s and 1960s was due to two factors. First was the discovery and exploitation of the unchoveta fishery of Peru; which sent Peru from having less than 1 percent of the world fisheries landing in 1950 to 17 percent in 1967 and a position us the top fishing nation in the world. The second factor was the introduction of modern fish-capturing and handling gear, including factory ships, larger nets, and more accurate methods of locating fishes. Most of the effort was expended on the continental shelf regions around the world with the result that at the present time, perhaps most of this limited area has been fully exploited for its existing fish stocks and many have been over fished. It would therefore appear that little increase can be expected from these areas. What about the remaining 93 percent of the world's oceans?

Currently the major oceanic fisheries of the world are for various tunas. Swordfish bill fish and so forth. In 1977 the landings of this group were 2.3 million metric tons or only 3 percent of total fishery landings for 1977. However, this 2.3 million to represented a significant increased in landing if over previous years and is probably partly due to the change from long line and pole fishing to the advent of purse seines (Table 5.1) Whereas 15 years ago the tuna stocks were considered nearly impossible to over fish the advent of purse seines dramatically changed that fact in a few years.

Today there is considerable concern about the continued ability of these fish stocks to sustain the current fishing pressure. Bottom fishing on the abyssal plains under the oceanic area is not feasible, since the diminished amount of energy reaching the abyssal floor would not allow the existence of substantial fish stocks. Furthermore, as we noted, the surface waters of these open ocean areas are impoverished with respect to nutrients and phytoplankton. They cannot sustain the populations of fishes, even in the surface waters, that the continental shelf waters are able to. They are "watery deserts." This suggests that, even through this is a vast area, it cannot match the production of the shallow coastal waters. In fact, we are approaching the limit of 'what we can obtain from the sea. What, then, is a reasonable estimate for the *maximum sustainable yield* of food that we can expect from the seas? By maximum sustainable yield, we man the largest number of fishes that can be harvested year after year without diminishing the stocks.

Perhaps the best estimate is based upon total primary productivity of the oceans. As we have already discussed, all food chains and tropic levels are ultimately dependent upon photosynthesis. The energy fixed in photosynthesis is finite in the world's oceans. By knowing the trophic level of the various fishes taken as food and the efficiency of energy transfer from level to level, it is thus possible to obtain a figure for maximum production. Such was the approach taken by Ryther (1969) to estimate potential yield. In this estimate, he first divided the ocean into three areas, each representing waters with significantly different levels of primary productivity.

The largest area (90 percent) was that of the Open oceans of the world, but it had the lowest primary productivity (Table 5.2). The second area in size (9.9 percent) was the neritic zone, and the final–highest in productivity but smallest in size (0.1 percent)–was the restricted area where upwelling occurs (Table 5.2). Next, he estimated the efficiency of energy transfer among trophic levels in each division and- finally, the trophic level of the major fishes captured in each (Table 5.3). Combining the data, he arrived at a total world fish production of 24 x 107 tons (=240 million tons).

Table 5.2. Division of the Oceans into Provinces based on Level of Primary Productivity.

Province	*Percentage of the Ocean*	*Area (Km^2)*	*Mean Productivity (grams of carbon per m^2 per year)*	*Total Productvity (10^9 tons of carbon per year)*
Open ocean	90.0	326×10^6	50	16.3
Coastal zone	9.9	36×10^6	100	3.6
Upwelling areas	0.1	3.6×10^5	300	0.1
Total				20.0

This figure, however, does not represent the total amount that may be taken. A significant fraction of this 240 million tons must be left as breeding stock to sustain future yields. Secondly, another large fraction is consumed by other carnivores in the sea. Ryther estimated these latter fractions to account for somewhat more than half of the total production, thus leaving about 100 million tons as the sustained, yield for the world's oceans. Other scientists, such as Cushing 1969 and Shaefer (1965), have also estimated the total harvestable amounts of fishes in the sea. These estimates have been made on the basis of primary productivity, on the basis of extrapolating from current landings, and on the basis of trophic level and ecological efficiencies.

Table 5.3. Estimated Total Fish Production of the World's Oceans Based on Three Provinces of Table 5.2

Province	*Primary Production (tons of organic carbon)*	*Trophic Levels*	*Efficiency*	*Fish Production (tons, fresh weight)*
Oceanic	16.3×10^9	5	10	16×10^5
Coastal	3.6×10^9	3	15	12×10^7
Upwelling	0.1×10^9	1½	20	12×10^7
Total				24×10^7

Because scientists disagree on the proper figures to use for primary productivity, ecological efficiency, and average trophic level, these estimates have varied from a low of 60 million tons to a high of 290 million tons. Given the current world marine fisheries situation in which, despite intensive effort and the wide use of modern gear, the catch appears to be leveling off in the vicinity of 60-65 million metric tons, it would seem prudent to accept limits somewhere around 100 million metric tons as the maximum sustainable yield. That we may be reaching the limit on the number of fishes that may be taken is manifested in the decline of various fisheries.

Overexploitation

In recent years, there have been abundant example of the decline of fish stocks of all types in all areas of the world. Some of these declines are clearly due to instances of overexploitation, but others have more complex origins. All indicate that we are approaching or exceeding the maximum amount of fishes we can take from the oceans, but also that we have often mismanaged many fisheries so that we have not achieved a maximum sustainable yield. The signs of over fishing are normally a decline in average size of fishes and an increase in effort needed to land the same amount of fishes. Age structure also changes. In some cases over exploitation has been compounded by certain natural changes in the environment that also acted to reduce the stocks.

In most situations, however, over exploitation is the direct result of human activity. Three recent examples are discussed here, one with environmental consequences affecting the fishery, one in which causes are uncertain, and one related directly to humans. The small clupeid fish known as the Peruvian anchoveta (*Engraulis ringens*) is a small filter feeder that occurs in large schools in the upwelling water mass of the Humbolt Current off the west coast of South America, where it feeds on the abundant plankton. This small fish occurred in tremendous numbers in these cold waters before the advent of human fishing.

The major predators were various species of marine birds. Collectively called "guano birds", which existed by the millions along these shores and produced huge quantities of droppings or guano, which were harvested for fertilizer. The anchoveta population depends on an abundant plank ton, which, in turn, depends on the nutrient-rich upwelling water. In certain years the upwelling is blocked in this area by an influx of warm surface water known, as the EI Nino Current. When this occurs, plankton disappears and the population of anchoveta drops as do the bird numbers. Hence, there is a natural abundance cycle. Direct human exploitation of the anchoveta began in the 19508 and the fishery increased at a phenomenal rate until the late 1960s. The catch was 10 million metric tons or more than one-sixth of the entire world's catch!

This huge catch was accompanied by a decline in bird numbers. Since they had fewer fishes to eat. The fishery experienced one drop in numbers in 1965 as a result of the occurrence of the EI Nino, but this temporary setback was made up the following year. Fishery experts had estimated a maximum sustainable enchoveta yield of 9.5 million metric tons. This was exceeded in 1967 when over 10 million tons were landed. By 1970, over 12.3 million metric tons were landed, but the size of the fishes was decreasing, and it was apparent to experts that the fishery was in trouble and that too few fishes were escaping to provide the breeding stock.

As could have been forecast, the crash came in 1972, aided by another EI Nino, so that landings dropped to 4.7 million tons. This was followed by an even worse year in 1973, when only 2.3

million tons were landed. Subsequently, this fishery has not recovered, so that even in 1977 only 2.5 million tons were landed. Thus a once huge resource has, due to over fishing been reduced far below its sustainable yield, and the human population of the world has been denied a significant protein supply. Much more mysterious is the demise of the California sardine fishery.

The California sardine, *Sardinops caerulea,* is also a small, plankton-feeding clupeid fish wish a range from Mexico to southern Alaska. In the 1930s, this species was the basis of the greatest fishery in North America. During the 1936-1937 season, fishermen landed 650,000 tons in California alone, the major fishing region. The declare began in the 1940s, when the age structure of captured sardines began to change to older age groups, reflecting the decimation of the younger fishes and their failure to recruit to the fishery. Despite the establishment of the California Cooperative Oceanic Fisheries Research Program in 1949, the fishery was commercially dead by the early 1950s.

In Monterey, California, a centre of the industry, the famed Cannery Row of Steinbeck closed up. The fishery has not recovered in the nearly 30 years since then, though a few tons of sardines are landed as "incidental", captures each year. Although most people attributed the demise of the fishery over the to-year period from 1940 to 1950 to over fishing, it seems apparent now that in this case, other factors were also at work. The decline of the fishery began with the onset of a 15-year period to unusually cold water in the sardine habitat. It is felt by Marr (1960) that this cold water affected the sardine population adversely, either directly through a reduction in breeding or through an effect on the food supply. The result was a catastrophic decline in numbers.

If over fishing were the only reason for the demise, one would have expected that in the 25 years during which there has been no fishing, the stocks would have recovered somewhat or at least returned to their known breeding grounds; but this has not occurred hence, the strong suspicion that other factors are involved. Finally, Soutar and Isaacs (1969) have presented evidence from the examination of the scales of sardines preserved in layers of sediments in certain anoxic basins, which the California sardine naturally undergoes huge cyclic changes in population numbers, cycles which range from 500 to 1700 years. If this is correct, the California sardine would be present in numbers for a certain period of years, and then vanish for a longer period before reappearing.

During its absence, the niche is filled by another clupeid, the anchovy. Currently, the increase in anchovy numbers would seem to substantiate this theory. If true, Cannery Row might again bustle to the activities of fish rendering in only 500 years or so! The final example of the demise of a fishery is much simpler, for it is the direct result of human over fishing. Here, we refer to the whaling industry. The ignominious end of the world's whale fishery will undoubtedly be observed in the next few years. In contrast to most animals taken in various fisheries, whales are warm blooded mammals which have a long life span and a very low reproductive rate. This means that they are very slow to build up their numbers once reduced and are easily reduced to extinction.

Large-scale whaling began in the North Atlantic in the sixteenth century. At first, methods were primitive and refuses from whalers existed for most species. However, the advent of long range vessels, and especially the harpoon gun with explosive charge. Spelled the end of whale stocks in the North Atlantic and North Pacific by 1900. Attention then turned to the large stocks of baleen whales in the Antarctic region, where most whaling of the twentieth century has been concentrated.

Coincident with the exploitation of these southern stocks was the advent of huge factory ships, more speedy catcher boats, and airplanes and helicopters to find the whales. All these technological advancements led to a greater slaughter of whales in the first 60 years of this century than had been achieved by all whaling efforts in all preceding centuries. In about 60 years of Antarctic whaling, Rounsefell (1975) notes that perhaps 1,300,000 whales were taken! In an attempt to regulate this fishery, the International Whaling commission (IWC) was set up in 1948. It as to obtain data on whale stocks and to set catch quotas for the various species so as to conserve the stocks. Unfortunately, because the Commission had no enforcement powers, because it had no independent means of obtaining data on the status of the various whale stocks outside of what the whalers reported, and because it was subject to intense political pressure by the whaling nations which it comprised, its history of attempts to conserve stocks has been a mockery.

It has consistently set its quotas too high and refused to heed the advice of its own and outside experts. As a result, the stocks of whales have continued to decline to the point where today many species are on the verge of extinction, and others are so reduced in numbers they are commercially extinct. Along with the demise of the whales has been the demise or the whaling industry. When the IWC was set up in 1948, its 20 member nations were whalers. By 1979. Only two remained as whalers, Russia and Japan. The dramatic decline in this fishery can be understood from a few examples. In the 1930-1931 season in the Antarctic, 25,000 blue whales *Balaenoptera musculus)* were taken, but by the 1963.1964 season, only 112 were taken. In 1968, the IWC reluctantly gave complete protection to the species. Declining blue whale stocks forced whalers to change to fin whales *Balaemoptera physalus)* after World War II, and from 1955 to 1961, more than 25,000 were taken each season. By 1965, this number had dropped to less than 2,500 and by 1974. 1975 to less than 1,000. Sei whales *Balaemoptena borealis)* went from 25,453 in 1964-1965 to 3,859 in 1974-1975. Clearly, the unrealistically high quotas set by the IWC and the greed of the whaling nations have reduced the whale stocks to the point where it is, or soon will be, uneconomical to continue whaling.

Even if all whaling is stopped now as many countries have urged, it will take many species years, perhaps centuries to recovery because of their low fecundity and the long time necessary for them to reach reproducing size. For some species such as the blue whale, McVay (1966) has suggested that the numbers may now be so low that the remaining few will be unable to find mates in the vast .reaches of the ocean. Extinction may be inevitable. That there is some hope for recovery of at least some of these leviathans lies in the recovery of the California gray whale *(Eschrichtins gibbosus)*. This species was reduced to only a few individuals in the early decades of this century; but since complete protection in 1938, they have now increased in numbers to more than 8,000.

Tragedy of the Commons

In 1968, Garrett Hardin published an article in *Science* magazine entitled "The Tragedy of the Commons." In this article, he pointed out that freedom of access to a common resource ultimately brings ruin to all who enter. The core to this argument is that whenever there is a resource held in common by a large number of individuals or political entities, such as countries, each entity will try to maximize its own gain from that resource because not to do so will allow another to do so and because the negative aspects of acting selfishly do Dot devolve on the one acting alone, but on all who share the resource. Unfortunately, such an analysis is made by each and every entity sharing the resource. As a result, the resource is destroyed. Therein is tragedy. The plight of the whale fishery can certainly be likened to such a tragedy. It may also be applicable to other open-ocean fisheries, such as tuna, or potential fisheries, such as krill.

Open-ocean fisheries lie in international waters-in other words, beyond the control of anyone country-end presumably are held in "common" for the good of the world's people. Since they are owned by no one, they and their resources are open to access by all. Therefore, those countries with the means to exploit these areas do so because it serves their own interests. They do not have to answer to another nation or regulatory agency, and the tragedy is that, if they don't catch the fishes, another nation will, it is just this short-term greed that has destroyed the whale fishery and may next do so for tuna, Recent interest in mining of deep ocean manganese nodules may also soon fall into this category.

Regulation

The decline of many world fisheries due to the tragedy of common access and the pressures of an increased demand for food by an ever-increasing human population have led to friction among fishing nations and various attempts at the regulation of fisheries resources. As long as the fishery exists on the continental shelf where individual nations have jurisdiction by international convention, the fishery can be controlled by a single nation. In the past, the area of an individual nation's jurisdiction was 12 miles (19.3 km) in recent years; however several countries have unilaterally acted to extend these limits out to 50 or 200 miles. This has caused several confrontations such as the "cod war" between Iceland and Britain and the seizure and fining of many United States tuna boats by Ecuador and Peru.

Most recently, through the actions of the United Nations, a standard 200-mile wide fishing and economic zone was established for, the waters of each coastal nation. This means that individual nations will now have full control of all fishery activity within 200 miles of their shores. International agreements among nations for the control of fisheries have proven to be more difficult to bring about and also to make work effectively, as witness the IWC. However, there are some outstanding examples that have worked. One of the oldest is the International fur Seal Treaty, established in 1911 among the United States, Russia, Japan, and Canada, to regulate exploit ion of the northern fur seal *(Callorhinus ursinus)* It has been an outstanding success in regulating fur seal numbers to achieve a sustained yield.

Another success has been the International Halibut Commission, established in 1924 between Canada and the United States to regulate the then declining Pacific halibut fishery. This was the first fish species to be conserved internationally. After establishment of the Commission, the stocks of halibut recovered (Figure 7.2). Since these early attempts, other international, regulatory agencies have been established, including the Atlantic and Baltic Commissions, The International Tuna Commission, and the International North Pacific Fisheries Commission (INPFC). Many of these more recent regulatory agencies have had a beneficial effect on certain fisheries, but problems still remain. The major, one is various commissions have limited jurisdiction geographically or are limited to only certain species, Thus, the waters around Japan are not in the jurisdiction of the INPFC and the Halibut commission regulates only that species. In order to effect a truly worldwide regulation of fisheries, the current commissions or others will have to expand their jurisdiction to include all species in their areas, and those areas, mainly in the Southern Hemisphere, where no regulatory agencies exist will have to be brought under control.

New Fisheries

Given the current leveling off of the world fisheries catch and the decline or plateauing in many long-established fisheries are there any prospects for finding substantial new fisheries? The answer appears to be yes, but whether these new fisheries will be able to contribute significantly to the human food supply is much less certain. Now that the stocks of baleen whales have been reduced to the point or extinction in Antarctic waters, a new fishery has developed for the "krill" (euphausiids) that formerly constituted the food of the whale. Krill exist in untold numbers in the seas around Antarctica, and they are large enough (2 inches) so they may be captured without resorting to very small-mesh plankton nets. The most conservative estimates of stocks of krill are 50 million tons, but other estimates range up to 100 million tons, according to Idyll (1970). Perhaps as many as 50 million tons could be harvested annually.

Thus, the potential for this fishery is enormous, certainly the largest of any unexploited fishery in the wc today. However,

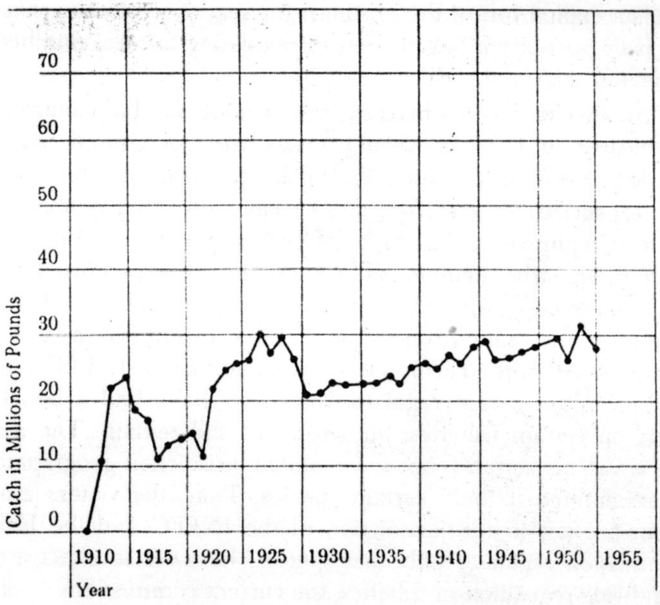

Fig. 5.3. Exploitation and recovery of the Pacific halibut fishery.

much remains to be learned about this fishery. Another area of potential exploitation, as yet relatively untapped, is the mesopelagic zone, especially off the continental slopes in 300-1000 m. Recently, substantial stocks of blue whiting have been discovered in this zone off the British Isles. This zone also contains substantial numbers of squids and small mid water fishes such as lantern fish. Finally, the entire Indian Ocean has a low level of fisheries development by current standards. It produces only about one fifth the catch per unit area of the Atlantic and Pacific. It would appear that increased efforts in these waters would increase the catches without over exploiting the stocks.

On the basis of primary production alone, Idyll (1970) estimated that this ocean can produce 10 million metric tons of fisher as opposed to a little over 4 million tons in 1974. We can summarize this section by saying that current world fisheries appear to be exploited to the maximum, with: many showing signs of definite overexploitation due to unregulated access and over fishing. Furthermore, there appear to be only limited amounts of increases to be expected from fishing stock and underexploited areas, with the possible exception of the Antarctic krill. The estimate of a total sustainable yield of 100 million metric tons or less for the world's oceans appears reasonable. With catches now reaching 64 million tons can expect little significant increase in food from the sea. Therefore, the Seas cannot and will not be the source of substantial amounts of food for an increasing human population.

Mariculture

As the limit is in sight for the harvesting of wild populations of marine animals and plants in fisheries, humans have begun to work with culturing various marine and estuarine species in situations that are more analogous to our terrestrial based agriculture, a field that is termed *aquaculture* Aqua Culture seeks to rear selected aquatic plants and animals under con trolled conditions so as to increase the amount of food available to humans with respect to that obtained through traditional fishing. We are concerned here only with marine aquaculture or *marl culture.*

History and Extend

Despite the evidence of the occurrence of culturing of marine organisms back at least as far as ancient Rome aquaculture through the centuries has never become a significant contributor to human food. Food from the seas remained the province of the hunters and gatherers, the fishermen. Only in this century has aquaculture received wide recognition and undergone significant development. Partly, this development is in response to the pressure for more human food, especially protein and partly from the declining wild stocks. Aquaculture encompasses marine organisms raised only for human food, but also for other products, such as pearls, and terrestrial animal food. Aquaculture is practiced throughout the world in the tropical and temperate zones, but is more technologically advanced in the industrialized nations of the Northern Hemisphere. At present, it is confined to shallow coastal embayments or artificial ponds.

Species Cultivated

Few marine organisms are currently cultivated and most are either the type that fetch a premium price on the market, which economically justifies the high cost of their culture, or else the type that produce large amounts of biomass under intensive culture. In the former category, we could include certain species of shrimps and abalone cultured in Japan. In the latter category are oysters and mussels. As might be anticipated, estuarine and bottom-dwelling species are more commonly cultivated than pelagic, open-ocean species. Fishes that are successfully cultured commercially include several species of Pacific salmon. *(Oncorhynchus), mullets (Mugil, Chanos),* yellowtail *(Seriola),* and flatfishes (*Pleuronectes, Solea*) Crustaceans cultured for mark species of shrimps and prawns *(Penaeus, Metapenaeus, Leander).* Among the molluscs, oysters *(Crassostrea Ostrea* and mussels *(Mytilus)* are subject of intensive commercial mollusks raised include several clams *(Anadara, Tapes, Mactra, Mercenaria, Meretrix)* and the abalone *(Haliotis)* the total list of' fishes, crustaceans, and mollusks taken for food in various fisheries, this is a very short list. Other species have been cultured experimentally, including lobsters, scallops, and crabs but at present, no commercial industry exists. This suggests that there is considerable room for future expansion.

Yields

Even though aquaculture has lagged far behind agriculture in development, its potential for production of human food even in its current low state is considerable. For example, Bardach (1968) has noted that the unmanaged common oyster grounds along the last coast of the United States yield only about 10 kg of oysters per hectare, while the intensive culture of oysters in hanging racks in Japan yield 58,000 kg/hectare. Even larger yields are possible with mussels. In Spain an intensive culture on hanging racks produces 300,000 kg/hectare. A similar situation occurs with fishes. In Indonesia, the milk fish *(Chants chanos)* is cultured in brackish ponds, but is not subjected to such intensive care as in Taiwan. Yield in Indonesia is 400 kg/hectare, but 2,000 hectare in Taiwan.

Presently, the world yield from aqua cultural practices is probably over 4 million tons, but most is from fresh water. Thus, mariculture currently contributes little human food yield. However, the science of mariculture is as yet at a low level of development. If the above examples are any indication of what can be expected, it would appear that the potential future Yield could be substantial however; it is not great for reasons discussed below.

Problems and Restrictions to Mariculture Development

Mariculture is not a panacea for the problem of feeding the human race, nor is it likely to contribute significantly to food supplies in the immediate future. The reasons for this are numerous space permits mentioning only the more significant. Marine animals require, first of all that the water in which they live has the proper physical and chemical characteristics. This is difficult to maintain in ponds or tanks and requires complex, filtering and water treatment to remove wastes and other toxic materials. Thus, attempts to raise marine organisms on a large scale mean an often economically crippling investment in equipment to maintain the water quality. Another drawback is that many marine species go through a series of larval stages, each requiring different conditions and food, before attaining adult size.

To successfully rear these forms is often too costly or simply impossible in captivity. Still other species are pelagic in the open ocean and will not survive when confided in smaller spaces. Since we cannot control animals in the pelagic areas, we cannot culture there. Finally, there is the problem of various diseases and parasites. Which tend to proliferate under captive or crowded conditions? What this means then, is that present and future mariculture appears most practical for those species with fewer or no larval stages, those that are nonpelagic, and those that live and reproduce well in crowded conditions.

Furthermore, our inability to control conditions or organisms in the open ocean means that future mariculture and increased yields must come from sheltered inshore bays, lagoons, and estuaries. Since the areas suitable for such culturing are a very minor part of the total world oceans, even at maximum culturing intensity, we cannot expect really substantial quantities of food from mini culture in the near future.

Pollution

The seas have been considered as the ultimate appropriate dumping grounds for the wastes of human societies, in the same way that fishery stocks were one considered inexhaustible, so we have felt until recently that the immense volume of the world's oceans had an infinite capacity for absorbing all of our waste materials. Dilution, in other words, was the solution to pollution. In recent years we have come to realize that however large, the oceans are not infinite in their capacity to absorb wastes and that certain of our wastes in very small amounts may have significant effects on communities and species. It is the purpose of this section to briefly discuss a few of the more significant sources of marine pollution and their effects or potential effects on marine ecosystems.

For more extensive information on pollution, the student is referred to the references at the end of the chapter.

Oil

Extensive media coverage over the last 15 years has certainly made the oil pollution problem in the seas widely known to the public. Oil pollution in the seas results primarily from the spillage of crude oil from offshore drilling platforms or from accidents involving tankers. What is less well known is that there are a few marine environments subject to natural oil contamination (Coal Oil Point in Santa Barbara, California, for example). Crude oil released into the sea usually floats, although some components .may sink, and after a time evaporation of certain fractions may cause even more to sink. Away from land, the floating oil probably has little effect on the majority of plank tonic and nektonic organisms.

The exceptions are the marine birds and certain mammals such as sea otters. When birds contact the oil, they become coated, and their feathers lose their insulating qualities. As a result, most will die of exposure in the winter or may be unable to feed, with the same result. Oil is therefore devastating to marine birds causing significant losses. Inshore, the oil may coat the shallow sub tidal arid intertidal and smother the communities. The devastation is nearly complete initially, but recovery usually occurs with time.

More serious than the all itself have been the various chemicals such as detergents used to break up or disperse the oil in the water. As Nelson-Smith (1973) notes, these were shown in *the Torrey Canyon* disaster in 1967 to have caused more mortality of marine organisms than the oil itself. Oil disasters have become more numerous with the increased oil demand of the industrialized world, with the necessity of transport of the oil from distant sources, and with increased numbers of offshore drilling rings. The immensity of oil disasters involving the giant supertankers is revealed by the sin king of the *Amoco Cadiz* off the Brittany coast of France in March 1978. In this case. 220,000 tons of oil spilled and one-third reached shore to contaminate 320 km (192 miles) of shoreline.

The spill contaminated or destroyed major commercial marine products, including oyster beds, mussel beds, and lobster holding pens. As of this writing, it is not known how long it with take the area to recover. Until recently, we had experienced no such extensive spills from offshore wells, though they are potentially more devastating, since the amount of oil gushing out is not limited by the hold of a ship. Such a disaster happened in the Gulf of Mexico where, in 1979, the Ixtoc I well blew out and spewed out 140 million gallons of oil in 295 days before being capped. It was the world's greatest oil spill. Only a few of these larger disasters per year could add up to significant destruction particularly of shallow water communities. We can expect more such disasters in future years with increased oil shipping and increased offshore drilling.

Sewage and Garbage

The discharge of human sewage and garbage into coastal waters is practiced throughout the world. The sewage mayor may not have had some treatment before discharge. Sewage adds a large volume of small particles to the water and large amounts of nutrients. In small volumes and with adequate diffusing pipes, it is difficult to detect any long-term effect on the communities if the open coast. In large volumes and in semi enclosed embayments, the effort can be devastating two examples should suffice. In southern California, the Los Angeles area discharges in excess of 330 million gallons of sewage per day at the White's Point Out fail off the Palos Verdes Peninsula. Extensive studies in and around this outfall and others in the area over a period of years by personnel of the Southern California Coastal water Research Project (Bascom, 1978) have revealed that sewage has caused significant degradation in benthic invertebrate communities in area near the outfall when compared to similar areas some distance away.

In addition, the kelp beds in the vicinity of the outfall have disappeared; sea archins have markedly increased; and diseased fishes are more prevalent. About 4.6 percent or 168 sq. km of the 3,640 sq. km southern California mainland shelf has been changed or degraded as a result of sewage discharge from four major outfalls. Kaneohe Bay on the bland of Oahu is a large, shallow embayment with restricted water circulation that was once known for its flourishing coral reefs. Since World War II, Banner (1974) reports that the area around the bay has been subjected to a tenfold increase in population. As a direct result of this urbanization, the bay has been subjected to massive domes tie sewage discharges along with significant siltation from runoff during storms.

The result has been the total destruction of the coral reef communities in two-thirds of the bay and their replacement by a massive, smothering growth of the green alga *Dictyosphaeria carvernosa.*

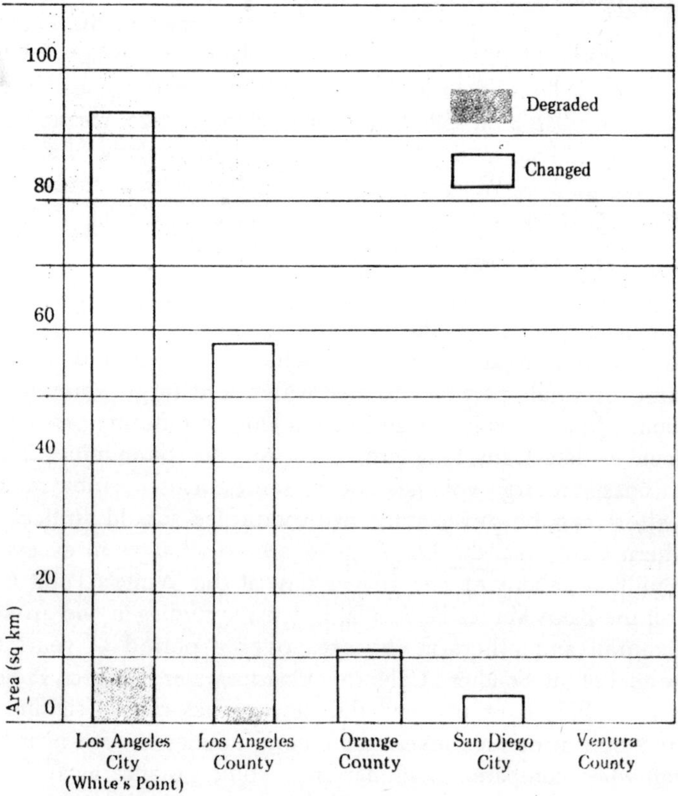

Fig. 5.4. Areas near southern California's major sewer outfalls that are "changed" or "degraded."

More recently, the sewage discharges have been eliminated from the bay and the bay appears to be recovering. In addition to sewage, large amounts of garbage are dumped into the sea each year from shore and from ships. What effect this material may have on the marine communities is as yet unknown. Evidence of the extent of the garbage pollution is found in the occurrence of small bits of plastic in plankton tows taken in many parts of the North Atlantic and Pacific oceans and in the digestive tracts of fishes, diving sea birds, and marine mammals. As yet, there is no evidence that these plastic bits cause mortality, though it has been suggested that they cause intestinal blockage. More ominous is that they do not decompose, and increasing use of plastics by modern human societies probably means increasing amounts in the oceans.

Chemicals

More insidious than oil or sewage, which are at least visible, are the various invisible toxic chemicals produced by the industrialized nations that find their way into the oceans ecosystems. These chemicals are often transferred through the food chains in the sea and exert their effects in animals and places removed in time and in space from the source. Because of this, it is difficult to pin down the effects of a given chemical, especially if the effects turn up years later. Certain marine organisms also enhance the toxic effects of many chemicals because they have the ability to accumulate the substances in their bodies far above that found in the surrounding water.

Another factor that tends to increase the effects of chemicals on living systems is *biological magnification,* in which the chemical increases in concentration in the bodies of organisms with succeeding tropical level. This results from the fact that these chemicals are not metabolized in an organism and therefore, the amount accumulated in the tissues remains their. When several such individuals are consumed by a carnivore of the next tropical level, the carnivore, in turn, gains the chemicals from all the individuals, which, in turn, increase the total concentration in its body.

Continuation of this process can lead to rather significant levels in the top carnivore, if the food chains are long. As we saw marine food chains are long and so top carnivores in marine systems of tell, have heavy loads. Humans also tend 'to consume marine organisms mainly from the higher trophic levels. This has already had an effect on certain of our fisheries such as swordfish and tuna, in which the levels of the element, mercury have often been higher than safe levels established by the FDA for humans, causing fishes to be rejected for consumption. Mort: tragic than that is the case of *Minimata disease,* documented by Goldberg (1974). In the late 1930s, the Chisso Corporation of Japan established a factory on the shoves of Minimata Bay to produce vinyl chloride and formaldehyde. By-products of this plant contained mercury and were discharged into the bay.

Through biomagnification, the marine fishes and shellfish accumulated high concentrations if the toxic compound methyl-mercury chloride. The fishes and shellfish were in turn consumed by the inhabitants of the area. About 15 years after the dumping of mercury into the bay began, a strange, permanently disabling neurological disorder began 10 appear among the inhabitants, especially among the children. The disorder was called Minimata disease. The cause was diagnosed as due to mercury poisoning in 1959, but it was not until the early 19608 that the active mercury compound was identified and the link to' the factory discharge established. It took even longer until the 1970s, for Japan to halt the discharge of mercury into the ocean and for other nations to blacklist mercury dumping and establish standards for acceptable levels in food. A second, well-documented example is the case of the pesticide *DDT.*

Dichlorodiphenyltrichloroethane (DDT) was the first of the synthetic pesticides of a class known as chlorinated hydrocarbons. It was put into worldwide use in 1945 and was hailed as a boon to mankind for its effectiveness in destroying a wide range of insect pests while remaining relatively nontoxic to humans. It has, however, 11 few attributes that were ultimately to bring about serious environmental consequences. First, it is a remarkably stable compound in natural systems. It or its first break-down product, DDE, persists for years just how long is unknown. It is relatively insoluble in water but .quite soluble in fats or lipids, and it adheres strongly to particles. Although DDT is not used in the marine environment, it has entered the marine food webs through runoff from land. Precipitation and dumping, as a result, by the 1960s DDT had been found in marine organisms as remote as Antarctica.

The insidious effects of DDT in marine food webs began to be apparent in the 1960s, when certain top carnivores were found to have high levels in their tissues. DDT enters the marine food chains at the level of plankton where the chemical adsorbs on the surface of the plankton organisms or is dissolved in the lipid. Since it is not metabolized, biomagnification concentrates it through succeeding trophic levels to the top carnivore. Concentrations of DDT in certain marine fish such as Marckerel were found to exceed the permissible *5* ppm established us safe for human consumption by the FDA and many catches had to be destroyed. The most serious effects, however, occurred among marine birds. In birds, once DDT reached certain high levels, it had a direct in interfering with the calcium deposition in the egg shell. As a result, thin eggs were laid, which broke open when the birds incubated them, the greatest amount of attention was focused on the brown pelican *(Pelicanus occidentails)* in which successful breeding virtually disappeared in the main rockeries on the island off southern California in the late 1960s and early 1970s. As Goldberg (1976) notes, this was traced directly to the high levels of DDT in their tissues. What was curious, here was that most of southern California, is urbanized and had not used much DDT, certainly not as much that used in agricultural areas. Yet, the waters around Los Angeles had concentration of DDT averaging 370 ppm as opposed to 1 ppm off Baja California. Where did it come from? The source was discovered to be the white's point sewer outfall.

It turned out that a single chemical plant that produced most of the worlds DDT was located in Los Angeles and was dumping was its waste its wastes into this outfall at the rate of 100 tons a year a figure ten times as high as the total amount of DDT carried per year by the Mississippi river into the Gulf of Mexico. Thus us, the decline of the brown pelican was linked through the marine food web to a toxic chemical discharged from a single source. Subsequent cessation of DDT dumping in Los Angeles has been followed by the slow recovery of the pelican. The above are only two examples of the problems of toxic chemicals. They emphasize the insidious nature of the often much delayed effects and the great difficulty of tracing the source of a problem through complex food webs, about which we know relatively little.. Since modern industrial states now produce thousands of actually or potentially toxic chemicals each years and many will enter the oceans, how many more of these disasters' must we look forwards to in the future.

Miscellaneous Pollution Problems

In order to provide docking facilities for commercial and pleasure vessels, humans have engaged in a great deal of dredging of estuaries and bays. Such activities destroy large areas of productive habitat for marine organisms and may have effects beyond the immediate area if nursery grounds for commercial species caught offshore are destroyed. Not only are Communities destroyed by physically being removed, but the increased load of silt suspended in the water reduces light and hence photosynthesis and clogs the feeding and respiratory surfaces of many invertebrates, leading to their deaths. Such activity has in the past destroyed many productive areas world wide, and the expanding human population suggests that we can look forward to even more destruction in the future. The sitting of power plants along the sea coasts to take advantage of the marine waters for cooling purposes has led to the perturbation of marine communities by heated water discharge. This *thermal pollution* has been little investigated for the marine environment, indeed; thus far, the effects would appear to be restricted to the communities immediately adjacent to be discharge.

The deleterious effects of thermal discharges are most pronounced in the tropics, where the organisms are living naturally at water temperatures that are close to their thermal maximum. It takes little increase in temperature to stress such communities. An example of the effects of a thermal discharge on tropical communities is described by Banus and Kolehmainen (1976) in Guayanilla on the south coast of Puerto Rico, where a power generating station discharges heated water is to a lagoon. The discharged water is 8-10°C higher than the ambient sea water and produces water temperatures in the lagoon of 39°C in the summer. This is high enough to effect reproduction of the dominant mangrove tree in the bay, and no young rooted and growing mangrove seedlings have been found in the lagoon. A final consideration concerns pollution by the introduction of *alien animals.*

The devastating consequence on native fauna and flora through the introduction of alien animals and plants is well known and documented for the terrestrial environment. We know for example, of the loss of the American chestnut through a disease imported from Europe and more recently, the demise of our native elm trees through Dutch elm disease. Equally dramatic have been the devastations of European rabbits on the flora and fauna of Australia. What about introductions in marine waters? Our knowledge of the effects of introduction of marine species is limited.

The major intentional introduction of marine species have been those of commercial or sports fishing value. Thus, the eastern *(Crassostrea virginica)* and Japanese *(Crasostrea gigas)* oysters were introduced into the bays and estuaries of the afore mentioned limitation for attached plants all digress of water motion enhance the exchange of nutrients and wastes. Although species of phytoplankton will populate only region with conditions to which they are adapted, factors other temperature, nutrients and light levels undoubtedly are important determining the species composition of phytoplankton population is understood of the mechanisms that live rise to an abundance of particular species under certain conditions.

Grazing herbivores may consume only a parts of the size and types of it allowing certain sizes and types to dominate temporarily Little is understood of the mechanisms hat give rise to an abundance of particular species under certain conditions. Chemical by products of certain species probably exclude certain other species. Often details of individual cell behaviour are probably also important in the introduction and success of a species in a particular area. In some cases we can glimpse what these mechanisms are for example, both the large diatoms and the larger flagellates can move at appreciable velocities through the water.

The diatoms commonly sink downward, whereas the flagellates actively swim upward toward light. These are probably patterns of behaviour primarily for increasing exchange. But the interaction of such unidirectional motions with random turbulence or systematic convective motions is not simple, as it is with an in a inactive particle Rather, we would expect diatoms to be statistically abundant upward moving water and to sink out of the near surface layers when turbulence or upward convection is low. Conversely, flagellates should be statistically more abundant down welling water and should concentrate near the surface In low turbulence and slow downward water motions. These effects seem to exist of some continental coasts in summer flagellates may eventually collect in high concentrations. As they begin to shade one to another from the light each individual struggles close to the lighted surface. Producing such high density that large areas of the water are turned red or brown by their pigment.

The concentration of flagellates in these "Red Tides" sometimes becomes too great for their own Survival several special of flagellates also become highly toxic as grow older. Thus they sometimes both produce and participate in as death of fish and invertebrates that has been on own to give rise to such a high yield of hydrogen sulfide as to blacken the white houses of coastal cities. Large diatom cells, on the other hand, sperd it disproportionately greater time in upward moving regions of the water and an unlimited time in any region where the upward motion about equals their own downward motion. Diatom cells are there statistically abundant in upwelling water, and the distribution of diatoms probably is often are flection of the trubutant-convective region of the water.

Sinking and the dependance of the larger diatomes on upward convection and turbulence for support aids items in reaching upwelling regions, where nutrients are high, it helps to emplain their dominance in such regions and such other feature of their distributions as their high proportion in rich ocean regions and their frequent inverse occurrence with flagellates. Differences in adaptations to the physical and chemical products, probably reinforce such relations. In some areas, such as parts of the equatorial current system and shallow seas. Where lateral and vertical circulation is rapid, the species composition of phytoplankton is perhaps more simply a result of the inherent ability of the species to grow, survive and reproduce under the local conditions of temperature, light, nutrients, competitors and herbivores.

Elsewhere second-order effects of the detailed cell behaviour often dominate. Those details of behaviour that give rise to concentration on any dimensional scale are particularly important to all subsequent steps in the food chain. The grazing of the phytoplankton it principally conducted by the herbivores members of the zooplankton. A heterogenous group of small animals that carry out several steps in the food web as herbivores, carnivores and detrital feeders. Among the important members of the zooplankton are the arthropods, animals with external skeletons that belong to the same broad groups as insects, crabs and shrimp.

The planktonic arthropods induce the abundant copepods, which are in a sense the making equivalent of insects. Copepods are represented in the sea by some 10,000 or more species that act not only as herbivores. Carnivores of detrital feeders but also as external Or even internal parasites two or three thousand of these species live in the open sea. Other important arthropods are the shrimp like euphausiids, the strongest vertical migrators of the zooplankton. They compose the vast shoals of krill that occur in high latitudes and that constitute one of the principal foods of the baleen whales.

The zooplankton also include the strange bristle-jawed chateognaths or, arrow worms, carnivores or mysterious origin and affinities know only in the marine environment. Widely distributed, and abundant. The chaetognaths are represented by a surprisingly mall number of species, perhaps, fewer than so. Larvae of many types, worms, medusae, ctenophores, gastropods, pteropods and heteropods, salps, unpigmented flagellates and many others are also important components of this milieu. Each with its own remarkably complex and often bizarre life history, behaviour and form.

The larger zooplankton are mainly carnivores, and those of herbivores habit are restricted to feeding on the larger plant cells.

Much of the blood supply, however exists in the form of very small particles such as the nanoplankton, and these appear to be available almost solely to microscopic creatures. Eyesight has developed in many minute animals to make possible selective capture. A variety of vebs, bristles, rakes, combs, cilia and other structures are found, and they are often sticky. Slickness allows the capture of food that is finer than the interspaces in the filtering structures, and it greatly, reduces the expenditure of energy. A few groups have developed extremely fine and apparently quite effective nets. One group that has accomplished this is the "Larvacea".

A larvacian produces and inhabits a complex external "House", much larger than its owner. That contains a system of very finely constructed nets through which the creature maintains a gentle flow. The larvacea have apparently solved the problem of energy loss in filtering by having proportionately large nets; fine strong threads and low rate of flow. The composition of zooplankton differs from place to place, day to night, season and year to year, yet most species are limited in distribution, and the members of planktonic communities commonly show a rather stable representation of the modes of life.

The zooplankton are, of course, faces with the necessity of maintaining breeding assemblages and like the phytoplankton. With the necessity of establishing a reinoculation of parent waters. In addition, their behaviour must lead to a correspondence with their food and to the pattern of large scale and small scale spottiness already imposed on the marine relean by the phytoplankton. The swimming powers of the larger zooplankton are quite adequate for finding local small-scale patcho of food. That this task is accomplished on a large scale is indirectly demonstrated by the observed correspondence between the quantities of zooplankton and the plant nutrients in the surface waters. There are many large and small puzzles in the distributions of zooplankton. As an ex-dense concentration of phytoplankton are often associated with low populations of zooplankton.

These are probably rapidly growing blooms that zooplankton have not yet invaded and grazed on, but it is not completely clear that this is so. Chemical repulsion may be involved. The concentration of larger zooplankton and small fishes in the surface layers is much greater at night than during the day, because of a group of strongly swimming members that share their time between the surface and the mesopelagic region. Many small zooplankton organisms also make a daily migration of some vertical extent. In addition to its primary purpose daily vertical migration undoubtedly served the (migration) migrating organisms in a number of other ways. It enables the creatures to adjust their mean temperature. So that by spending the days in cooler water the amount of food used during rest is reduced. Perhaps such processes of the rate of egg development are also been controlled by these tactics.

,Convincing arguments have also been presented to show that vertical migration serves to maintain species so that they will be more successful under many more conditions that if they lived solely in the surface layers, This migration must also play an important part in the distribution of many species. Interaction of the daily migrants with the water motion produced by daily land-sea-breeze alternation can old the migrants off store by a kind of "Rectification" of the oscillating water motion. More generally descent into the (water) lower layers increases the influence of submerged counter currents.

There by enhancing the opportunity to regions and hence to find high nutrient revels and associated high phytoplankton productivity. Even minor details of behaviour may strongly contribute to success. Migrants spend the day at a depth corresponding to relatively constant low light waves, where the moment of water commonly is different from that at the surface. Most of the members rise some what even no the passages of cloud shadow. The principal food supplies of the pelagic population are passed on in incremental steps and rapidly depleted quantity to the larger carnivores, zooplankton, then to small fishes and squids, and ultimately to the wide range of larger carnivores of the pelagic realm.

The pelagic region contains some of the largest and most superbly designed creatures even to inhabit this earth: The exquisitely constructed please tunas. The multicoloured dolphin fishes, capturers of flying fishes; the conversional purposes; the shallow and deep feeding swordfishes and toothed whales, and the greatest carnivores of all, the baleen whales and some plankton eating sharks. Whose prey are entire schools of krill or small fishes. Seals and sea-lions feed far into the pelagic realm. In concert with these great predators, large carnivorous sharks a wait injured prey Marine birds, some adapted to almost continuous pelagic life, consume surprising quantities of ocean food, diving, plunging skimming and gulping in pursuit. These larger creatures of the sea commonly move in schools, shoals, and herds. In addition to meeting the needs of mating such grouping is advantageous in both defensive and predatory strategy.

Small fishes of several species commonly school together. Diverse predators also form loosely cooperative groups, and many species of marine birds depend almost whooly on prey driven to the surface by submerged predators. At night, schools of prey and predators are almost always spectacularly illuminated by bioluminescence produces by the microscopic and larger plankton. The reason for the ubiquitous production of light by the microorganisms of the sea remains obscure, and suggested explanations are controversial.

It has been suggested that light is a kind of inadvertent by-product of life in transparent organisms. It has also been hypothesize that the mission of light on disturbance is advantageous to the plankton in making the predators of the plankton conspicuous to their predators. The fall out of organic material into the deep dimly lighted mid water supports a sparse population of fishes and invertebrates with in the mesopelagic and bathypelagic zones are found some of the most curious and bizarre creatures of this earth. These range from the highly developed and powerfully predators intruters, toothed whales and sword fishes, at the climax of the food chain to the remarkable squids, octopuses, euphausidis, lantern fishes, gulpers and angler fishes that in habit the bathypelagic region. In the mesopelagic region, where some sunlight penetrates, fishes are often counter-shaded, that is they are darker above and lighter below as are surface fishes.

Many of the creatures of this dimly lighted region participate in the daily migration. Swimming to the upper layers at evening like bath emerging from their caves. There are some much larger stronger and more active fishes and squids in this region, although they are not taken in trawls or seen from submersibles. Knowledge of their evidence comes mainly from specimens found in the stomach of sperm whales and sword, fish. There is evidence that the sperm whales posses highly developed long range hunting sonar. They may locate their prey over relatively great distances. Perhaps miles, from just such an extremely sparse population of active bathypelagic animals. Although many near-surface organisms ate luminescent. It is the bathypelagic region that bioluminescence has reached a surprising level of development. With at least two-thirds of the species producing light. Clearly bioluminescence can be valuable to higher organisms, and the creatures of the bathypelagic realm have developed light-producing organs and structures to a high degree. Strong fishes may confuse predators by "target alteration" effects or by producing residual images in the predators vision.

Some squids and shrimps are more direct and discharge luminous organs are arranged on some fishes so that they can be used to countershade their siltouettes against faint light coming from the surface. Lights may also be used for locating a mate. A problem of this vast, sparsely populated domain that has been solved by some angler fishes by the development of tiny males that live parasitically attached to their relatively huge mates. Light level in the bathypelagic region can be much lower. This is most probably the primary difference that accounts for the absence of bioluminescence in higher land animals add the richness of its development in the ocean forms.

There are some 2000 species of the larger invertebrates lenown to inhabit the bathypelagic zone, But only 4 few of these species appear to be widespread. The barriers to distribution in this widely interconnected mid-water region are not obvious. The floor of the deep sea constitutes an environment quite unlike the mid-water and surface environments. There are sites for the attachment of the larger invertebrates that filter detrius from the water. Among these animals are representatives of some of the earliest multicelled creature to exist on tile earth glass sponges, sea little–once thought to been long extinct–and lamp shells. Larger numbers of active fishes and other creatures are attracted to the bait almost immediately. It is prosaldy fuel that several rather independent branches of the food fuel coexist in support of the deep-bottom creatures; one the familiar rain of fine detitus, and the other the rare, widely separated falls of large food particles that ate in excess of the local feeding capacity of the broadly diffuse bathypelagic population.

Such falls would include dead whales, large sharks or other large fishes and fragments of these, the multitude of remnants that are left then predators attack a school of surface fish and now. Undoubtedly, garbage from ships and kills from underwater explosions. These sources result in an influx of high-grade food to the sea floor, and we would expect to find a population of active Creatures adapted to its prompt discovery and utilization. Other sources of food materials are braided into these two extremes of the abyssal food web.

There is the rather subtle downward diffusion of living and dead food that results initially from the daily vertical migration of small fishes and zooplankton near the surface. This migration appears to impress a sympathetic daily migration on the mid-water population down to great depths, far below the levels that light penetrates. Not only such vertical migration bring feeble bathypelagic creatures near the bottom but also it accelerates in it self the flux of dead food material to the bottom of the deep sea. Benthic animals are much more abundant in the shallower matter off continents, particularly offshore from large rivers. Here there is often not only a richer near-surface production and less hazardous journey of food to the sea floor but also a considerable input of food conveyed by rivers to the bottom. The deep slopes of river sediments wedges are arc typified by a comparatively rich population of burrowing and filtering animals that utilize this fine organic material. The shallow regions of such wedges are highly productive of active and often valuable marine organism. The bottom is much more variable than the mid-water zone is. There are as a result more environmental riches for an organism to occupy, and hence we see organisms that are of a wider range of from and habit.

Aside from the wide range of form and fraction the benthic environment elicits from its inhabitants, there arc more fundamental conditions that influence the nature and form of life there. For example -the dispersed food material setting from the upper layers become much concentrated against the sea floor. In the mid-water environment most creatures must move by their own energies to seek food, using their own food stores for this motion. On the bottom, however, substantial water currents arc present at all depths, and creatures can await the passage of their food. Here large organisms can grow by consuming microscopic or even submicroscopic food particles. Clams, scallups, mussels, tube worms, barnacles and a host of other creatures that inhabit this zone have developed a wide range of extremely effective filtering mechanisms. Although the benthic environment enables the creatures of the sea to develop and major branch of the food web that is emancipated from successive microscopic steps. This makes little difference to the food economy of the sea.

The sea is quite content with a large population of tiny organisms. From man's stand point, however, the shallow benthic environment is an unusually effective producer of larger creatures for his food, and he widely utilizes these resources. In short, their is, of course, much to learn about all marine life. The basic processes of the food web, productivity populations distributions and the mechanisms of reinoculation, and the effects of intervention into these processes, such as pollution, artificial upwilling, transplantation, cultivation and fisheries. To learn of these processes and effects we must understand the (structure) nature not only of strong simple actions. But also of weak complex interactions, since the forms of life or the success of-a species may be determined by extremely small second and their order effects. In a synopsis way we can also say that the possible benefits of broad marine-biological understanding are endless. Man's aesthetic, adventurous, recreational and practical proclivities can be richly served.

6

ACID RAIN

Increasing acidity in natural waters and soils has been becoming problem over extended areas of the world, particularly North Eastern America and North Western Europe. This acidity has been associated with the transport and subsequent deposition of sulphur-dioxide: nitrogen oxides and their acid oxidation products. The concentration levels of the transported gases have been not sufficient to damage the environment directly, it is their accumulation which causes the problem. This accumulation was first reported as increasing acidity in Swedish lakes and rivers, an acidity that could not be explained by reference to Sweden's sulphur dioxide emissions alone.

It was consequently proved that much of the material responsible got transported from the highly industrialized areas of the UK and Central Europe, and that this transport had been aided by the tall industrial stacks which had been used in these areas to avoid a local problem. Similar conclusion were drawn from Finland, Norway, Austria and Switzerland, all of which were found to be net importers of sulphur compounds, in the course of an OECD Study during 1973, 1974 and 1975

Meaning of Acid Rains

Acid rains means in common language the presence of excessive acids in rain water. It has been one of the effects of air pollution. Every source of energy that we use–be it coal, fuelwood or petroleum products, has sulphur and nitrogen. These two elements when burnt in the presence of atmospheric oxygen are converted into their respective oxides-sulphur dioxide and nitrogen dioxide,

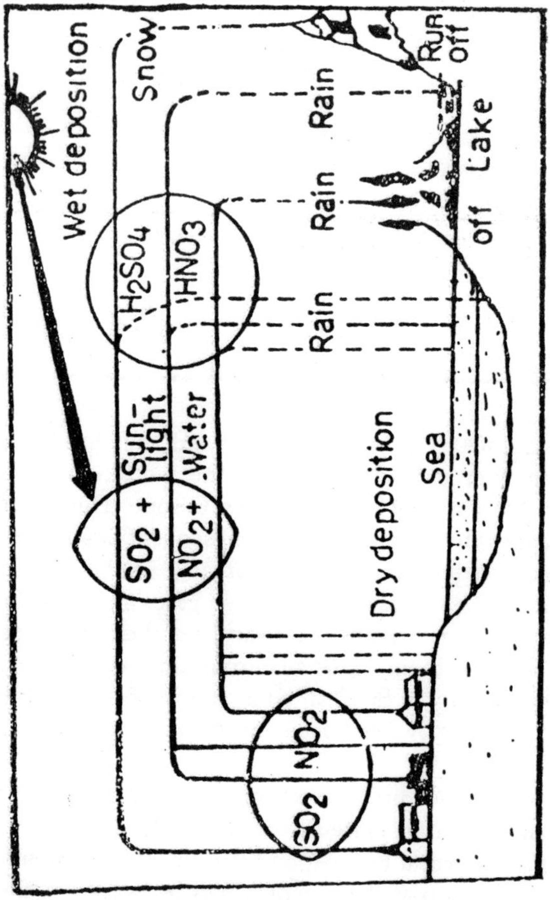

Fig. 6.1. Acid rain.

which are highly soluble in water. During rain these oxides react with large quantities of water vapour of the atmospheric to form acids like sulphuric acid, sulphurous acid, nitric acid and nitrous acid which then return to the earth's surface, with rain water or many remain in the atmosphere in clouds and logs.

Adverse Effects of Acid Rains

Probably the most worrying aspect has been that the area affected by acid precipitation has been increasing year by year, and its association with our energy consuming world leaves little prospect of alleviation, even with stricter emission controls. The repercussions on the environment have been difficult to quantify because they are a function of soil type and bedrock. It does seem, however, that the siliceous bedrock, thin soils and soft waters of the areas at present receiving the deposition have been particularly vulnerable. It gets related to the relative insolubility of the rock and consequent lack of dissolved species in the water which could founder (buffer) the acidity change. Significant reductions in fish populations have been observed, accompanied by an associated decrease in the variety of species at all levels in the food chain.

There are 15,000 fish less lakes in Sweden because of increased acidity, and some 100 such lakes in the Adironadak region of the USA. Soils in these regions similarly are having little buffering capacity and are subject to the leaching of minerals essential to plant growth. There have been soils which have been low in sulphur or high in carbonate, where an increase in acid sulphate could well be benign or even beneficial but these have been not predominant in North Western Europe or North Easter USA.

Acid rain causes a number of adverse implications. It tend to increase acidity in the soil, threatens human and aquatic life destroys forests and crops reducing agricultural productivity. Besides, it is able to corrode buildings, monuments, statues bridges, fences and railings, that costs the world 1450 million, dollars a year. At St. Paul's cathedral some stonework is being eaten away at the rate of an inch every 100 years. Even the British Parliament building has suffered serious damage from the presence of sulphuric acid in rain fall (The Times of India October 25, 1986). It creates a serious threat to human health also, since it contaminates not only the breathing air but also the drinking water and even food. The acids have been found to be very dangerous to the living organisms as they can destroy life.

Acid rain can play havoc with the human nervous system by making the person an easy prey to neurological diseases. This happens because these acids produce highly toxic compounds which contaminates the potable water and enter our body. Acid rain has been already an acute problem in North America and Europe. Crops and forests in Canada are being destroyed by acid rain due to pollutants emitted by industries in Northern USA. Acidity Kills fish, bacteria and algae and the aquatic ecosystem is destroyed.

Winds are known to carry air pollutants from one country to another. Air pollution in England now descends upon Sweden as acid rain. Acidification of soil changes its biology and chemistry. When the soil gets acidified, plants can absorb cadmium more easily and high levels of cadmium in plants has been dangerous for animals and human beings. Acid rain in Japan has damaged 5,000 sq. kms. of cedar trees in Kanto plain which lies to the North of Tokyo. This area is having high acid deposition brought about by air pollutants.

Acidic air pollutants have been responsible for many other-damaging effects like corrosion of metals, weakening or disintegration of textiles, paper and marble. Investigations are going on to know whether Taj Mahal is being affected due to pollutants released from Mathura refinery. Hydrogen Sulphide tarnishes silver and blackens leaded house paints, ozone produces cracks in rubber; its economic significance has been apparent. Air pollution may cause or contribute to a variety of safety hazards, e.g., hazards associated with reduced visibility due to smog etc. Air contaminants that have been detertable by the sense of sight, touch, smell or taste can be nuisance in many ways even if they do not result in direct adverse economics or health effects.

In the near future the developing countries like India will soon have to cope with this problem of acid pollution. Industrial areas with the pH value of rain below or close to the critical value had been recorded in Delhi, Nagpur, Pune, Bombay and Calcutta. Nadia acidity has been largely due to sulphur dioxides from coal fired power plants and petroleum refinery, emitting nearly 85 percent tonnes annually. The phenomenon of acid rain is becoming more and more common in Bombay with, several industries discharging sulphur oxides in the air making: rain water more acidic. The average pH value of acid rain at Calcutta has been 5.80, Hyderabad, 5.73, Madras, 5.85, Delhi. 6.21, and Bombay, 4.80. The situation in India has been likely to worsen in the near future with the increased tempo of thermal power plants.

7

Pelagic Food Webs

Epipelagic Food Webs

Deep-sea animals are most diverse under the surface waters of the subtropical and tropical belts. Here are clear blue waters, where photosynthesis by the microscopic algae of the plankton is possible down to a depth of 100 metres or more. Apart from the divergent regions of the equatorial current systems, where upwelled water brings nutrient salts to the surface and leads to relatively high levels of productivity, these warm central oceanic waters are low in nutrients and standing crops of plankton. Blue-green algae thrive in such surroundings and a dominant form, *Trichodesmium*, whose rust-coloured blooms may at times cover wide stretches of the warm ocean, is reported to use fixed molecular nitrogen for its growth.

In other respects the composition of the phytoplankton of the tropical and subtropical belts is much like that of other oceanic regions. But the size range of the algae extends below that of floras in nutrient-rich waters, such as are found in the antarctic and in upwelling waters off Peru and elsewhere. For comparable forms the smaller the cell the less are its nutrient requirements, but the more its surface for nutrient absorption compared to its volume. Thus, it somehow seems right that much of the flora of nutrient-poor tropical waters should consist of mobile µ-flagellates, such as cryptomonads and chrysophytes. To be but a few microns in size confers another advantage.

In subtropical and tropical regions in particular, it seems that the multiplication of the phytoplankton barely keeps pace with its consumption by all manner of herbivores in the zooplankton. One might then expect selection pressure to be high for means of countering the grazing inroads of zooplankters. To be small enough to pass through the meshes of a filtering system, particularly of copepods, is advantageous, though there is less escape from the mucus nets of the pelagic tunicates. But in overall terms of nutrition μ-flagellates may form no more than a small part of the grazing of herbivores. For instance, during an expedition to the eastern tropical Pacific counts of phytoplankton constituents were made (under an inverted microscope) from 200 samples covering 48 stations. When each count was expressed in terms of its carbon content, the main conclusion was that though flagellates (2-5μ in diameter) formed

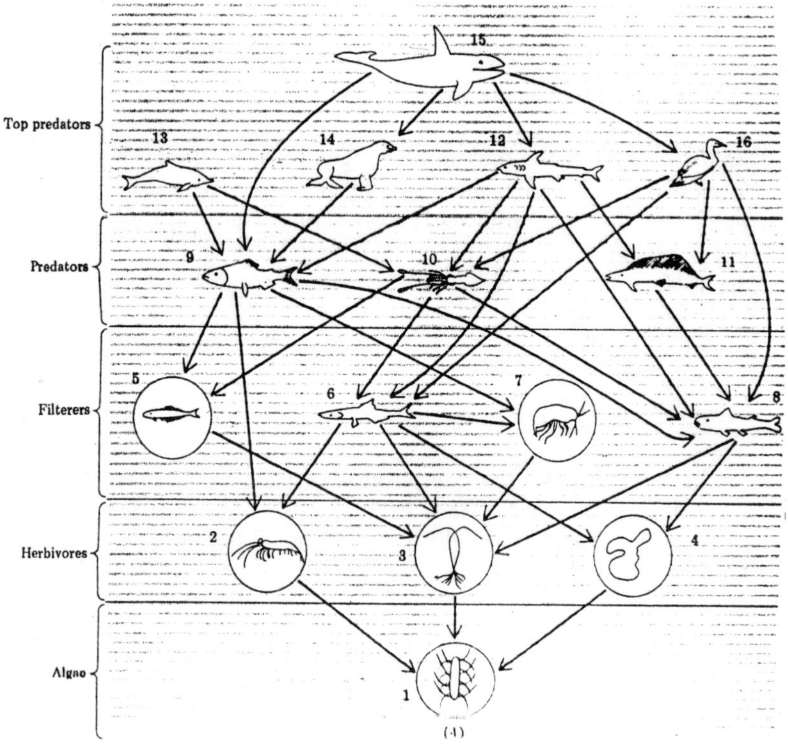

Fig. 7.1. Food webs and trophic structure of the pelagic community in three different geographical areas. Cold temperature waters. Algae: (1) diatoms. Herbivores: (2) euphausiids, (3) copepods, (4) pteropods. Filterers: (5) juvenile fish, (6) anchovy, (7) hyperiid amphipods, (8) lantern fish. Predators: (9) salmon, (10) squid, (11) lancetfish. Top predators: (12) mackerel shark, (13) porpoise, (14) seals and sea lions, (15) killer whale, (16) marine birds.

about 80 percent of the catches in numbers of individuals, they made only a minor contribution to the total content of phytoplankton carbon.

In equatorial waters most of this carbon was contained in the diatoms, whereas dinoflagellates were the main carbon contributors north and south of the equatorial region. No doubt the relative amounts of the major groups of algae change from time to time, and we must always remember that the fragile and elusive µ-flagellates are likely to be underestimated. There is no off-season in the production of phytoplankton in warm oceanic waters : throughout the year diverse zooplankters live on thinly dispersed crops particularly in the subtropical regions bounded by the gyres. In temperate and cold regions there is rich grazing in the favourable season, during which animals form stores of fats and oils to help tide them over the winter months. Whatever the place, there are times when the phytoplankton crops are so meagre and thinly spread that one may wonder how the herbivores manage to exist.

Indeed, such thoughts, stimulated by the quantitative studies of Hensen, Lohmann and Brandt towards measuring the productivity of the seas, led Putter to investigate the nutritional requirements of marine animals (as measured by their rates of carbon dioxide production). After comparing the metabolic needs of diverse species with ambient amounts of algal food, Putter concluded that marine animals must depend largely on dissolved organic matter, which they absorb through their integument. Krogh, who was drawn to this problem through his experiments to determine whether aquatic animals are able to absorb organic substances, first concentrated on an accurate method of measuring dissolved organic matter (DOM) in sea water. Though he found that there is much more DOM than particulate forms of organic matter in the ocean, he also observed that DOM is very largely resistant to bacterial decomposition.

Krogh concluded that the food of aquatic animals is generally in the form of organisms and organic detritus. Since the 1950s numerous experiments on diverse aquatic invertebrates have shown that most marine, but not freshwater, forms can take up glucose and amino-acids through their integument. But Korgh's conclusion still stands. Even the Pogonophora, which are confined to sediments rich in organic matter, have a restricted rate of absorption when in their tubes. Members of the micro-zooplankton, certain copepods and euphausids, shelled pteropods and pelagic tunicates form the main groups of oceanic herbivores. More than we realize, the lives of these diverse forms (and their predators depend on the natural fact that the phytoplankton is not randomly dispersed.

Statistical analyses of well planned series of samples show that the constituents of phytoplankton are 'overdispersed', with implies the aggregation of 'clumping' of individuals but not necessarily the forming of physically observable clusters. Studies of water column down to 150 metres off Bermuda, which was so homogeneous that measures of temperature and salinity varied by less than 0.05°C and 0.05 percent respectively, showed that the concentrations of phytoplankton species differed by several orders of magnitude at different depths. There are discrete layers of phytoplankton and at times such concentrations lie near the surface, at others near or even below the depth of the euphotic layer. Members of the micro-zooplankton aggregate in these layers.

In one way or another enough individuals of any herbivorous species seem to refute Putter's conjecture, which is also true of the largest marine animals. For instance, fin whales were estimated to need zooplankton concentrations in excess of 1.5 gm per cubic metre, whereas the average standing stock of zooplankton in the Subarctic Pacific is rather less than one-tenth of this amount. During crossings of this region Barraclough *et al.* used a 200 kHz echo-sounder and recorded shallow scattering layers that net hauls showed were composed almost entirely of copepod. *Calanus cristatus.* Such concentrations were considered to be more or less sufficients to satisfy the appetites of fin whales. While there is no off-season in the production of life in subtropical and tropical waters, there are seasonal changes. Indeed, monsoon comes from the Arabian word *mausium*, meaning a time or season.

During the northeast monsoon in the Indian Ocean measurements of carbon assimilated by the phytoplankton yielded a mean figure of 0.15 gmC per square metre per day, which is less than one-third of the production during the southwest monsoon. The invigorating effect of the southwest monsoon is widespread, but in the warm Pacific Ocean seasonal changes are most pronounced in equatorial regions. Here from July to December there is greater productivity than during the rest of the year, when there are northeasterly or variable winds.

In the great subtropical gyres, where the 'seasonal' thermocline is more or less permanent nearly every where there can be little

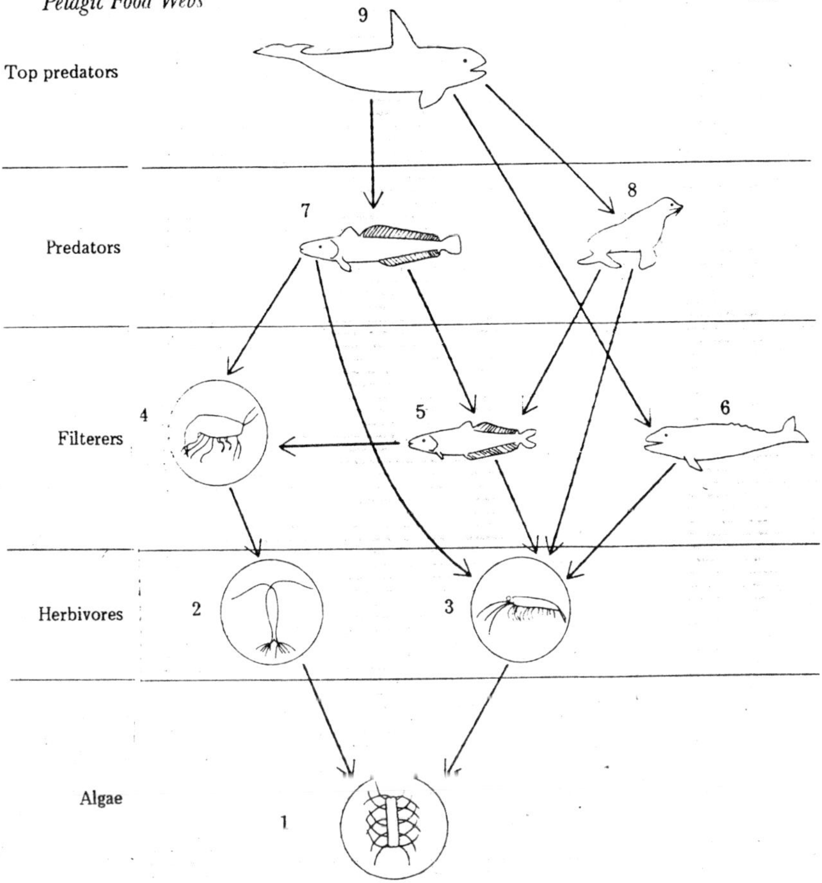

(*B*)

Fig. 7.2. Antarctic seas. Algae: (1) diatoms. Herbivores: (2) copepods, (3) euphasiids. Filterers: (4) hyperiid amphipods, (5) planktivorous fish, (6) baleen whales. Predators: (7) predatory fish, (8) seals and sea lions. Top predators: (9) killer whale.

seasonal change in the standing stocks of plankton. Marked seasonal changes in primary production no doubt lead to changes in the biomass and composition of the zooplankton. Concerning the productive zones of the equatorial Atlantic in the summer Bernikov *et al.* were impressed by the high biomass of plankton, the bulk of which consisted mainly of jelly-like animals, notably of pelagic tunicates. In the autumn most of the zooplankton consisted of crustaceans. Pelagic tunicates with their efficient plant-gathering nets of mucus and relatively little living tissue to maintain, seem well organized to respond quickly to an increase in primary productivity. Indeed, one salp with an instantaneous rate of population increase of 0.4 to 0.91 per day, can double its members in a single day. Swarms of this salp, which may be very widespread, are generated immediately after phytoplankton blooms, and thus are transient resources exploited to the full. Their small relatives, the appendicularians most of which live in subtropical and tropical regions, are even livelier opportunists. In the Black Sea the turnover time of Oikopleura is about three days.

Appendicularians are among the commonest groups of zooplankton (often second only to the copepods) and, like their relatives, are able to feed on the smallest kinds of phytoplankton. The mesh size of the filtering windows in their houses varies from species to species and when the filter becomes clogged with food, another house is quickly secreted (which may be several times in one day). In each discarbed house is a concentrated package of phytoplankton, detritus and mucus, food for other planktonic animals and bacteria. Appendicularians themselves are prey to such animals as jelly-fishes, siphonophores, arrowworms and lantern-fishes. By-products of zooplankters, particularly of appendicularians, salps and pteropods, form many of the flakes and larger organic aggregates of the 'marine snow' so often seen in the open ocean by divers and observers in submersibles. Alice Alldredge concludes, 'these large amorphous bits of mucus and detritus serve as surfaces on which many planktonic organisms can rest or feed. Bacteria and protozoans use them as permanent habitat.

Appendicularian houses and other particles of marine snow provide tiny solid substrates and introduce heterogeneity into an environment that is generally considered to be relatively homogeneous physically. Although these miniature habitats have not been completely explored, it is certain that they have influenced the adaptation of feeding strategies of many planktonic organisms'. Pelagic tunicates are what theoretical ecologists call *r*-strategists, opportunistic animals that grow and reproduce quickly when conditions become favourable. Indeed, nearly all gelatinous forms in the zooplankton, both herbivorous and carnivorous, seem to be *r*-strategists. Concerning the latter, medusae, siphonophores and crenophores evidently have a versatile adaptation system, which facilitates flexible synchronisation of the population booms with abundance of planktonic forms which they use for food. This synchronisation includes the capacity to vary the number of eggs; timing, rates and types of reproduction; to withstand long periods of starvation; to produce resting stages, and to form aggregations.

These properties provide passive pelagic predators with the capacity of "waiting" for a favourable biotic situation to occur'. At first sight, these gelatinous herbivores and carnivores seem to break the rule that *r*-strategists are small but if large in bulk, as most species are, they are small in protoplasmic content. To find small kinds of *r*-strategists, one need look no further than the protozoans

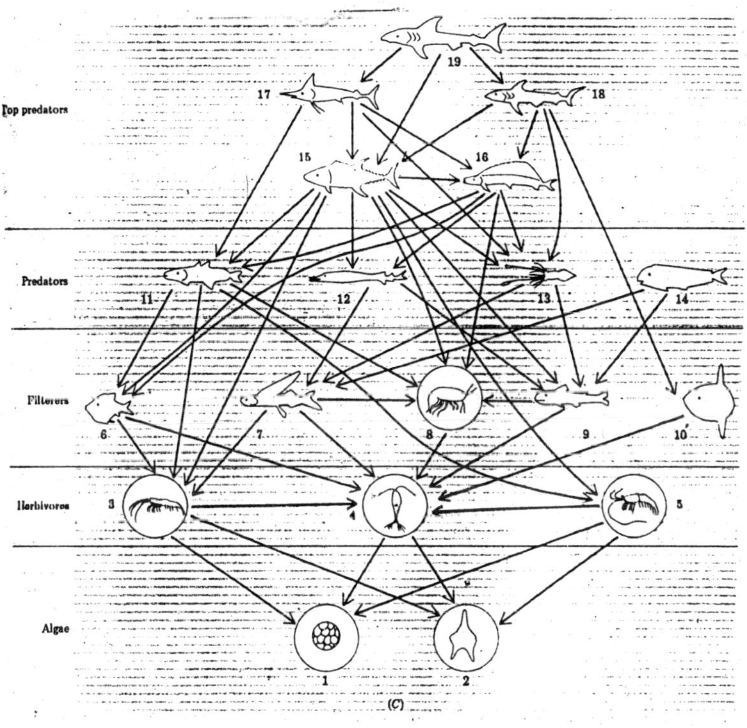

Fig. 7.3. Tropical seas. Algae: (1) coccolithophores, (2) dinoflagellates. Herbivores: (3) euphausiids, (4) copepods, (5) shrimp. Filterers: (6) vertically migrating mesopelagic fishes, (7) flying fishes, (8) hyperiid amphipods, (9) lanternfish, (10) ocean sunfish. Predators: (11) mesopelagic fishes, (12) snake mackerel, (13) squid, (14) dolphin (Coryphaena). Top predators: (15) tuna, (16) lancetfish, (17) marlin, (18) medium-sized sharks, (19) large sharks.

of the micro-zooplankton. Compared to those of cold regions, zooplankton communities of the warm ocean have a low biomass, a high diversity with no very dominant species, a higher proportion of carnivorous species, more species with a short life span and numerous herbivorous species with a reduced body size (but there is no such diminution in carnivorous species).

As Motoda and his colleagues say, one advantage of greater size in herbivorous zooplankters of cold waters is that they have extra space to store fats etc., to help them over the winter season. The smaller herbivorous species of the warm ocean live faster and breed faster but store little in the way of food reserves, for phytoplankton is always available, even if the supply may be meagre, as it most often in the deep-blue gyral deserts of the subtropical belts. In the North Pacific gyre productivity is generally limited by the quantity of nitrogenous nutrient present in the water. Indeed, nitrates are insignificant in quantity; the main source of nitrogen for plant growth seems to be ammonia and urea produced as waste products by the zooplankton and nekton. The impoverished nature of subtropical waters, compared to equatorial and more northerly waters of the western North Pacific, may be seen in Fig. which shows the geographical variations in phytoplankton crops and the biomass of zooplankton in the surface waters.

In subtropical regions, where the upper waters are kept stable but impoverished by a marked thermocline, plant production barely keeps pace with herbivore consumption : there is, as Cushing calls at a quasi-steady state production cycle. In the Indian Oceans, Timonin who analysed zooplankton communities, especially their copepod, euphasiid and chaetognath constituents, in specific and trophic structure, contrasts regions of stable water stratification with those, in the equatorial belt where current divergences lead to upwelling and greater productivity. Stable regions support communities with a low biomass but high species and tropic diversity.

Communities in region of low divergence were more or less intermediate in the values of the above ecological indices. After one organisms becomes the food of another, matter and energy are lost in the process that transform prey tissue into predator tissue. Any such transformation can be seen to have a certain transfer efficiency, which Ricker defines as '....the percentage of the prey's annual production that is incorporated into the body tissues of the consumer species. This, in turn, is compounded of two processes characterized by separate coefficients:

E, the ecotrophic coefficient–the fraction of a prey species annual production that is consumed by predators (trophic referring to nutritive or food levels):

K, the growth coefficient–the predators annual increment of weight divided by the quantity of food they have consumed.

Ricker suggests that an overall average growth coefficient for herbivores could hardly exceed 15 percent, though the figure may be around 20 percent for fish and other consumers of animals food. Concerning the ecotrophic coefficient of plant consumers a figure of 66 percent seems reasonable, while for secondary consumers, such as lantern-fishes which feed on zooplankton, 73 percent in proposed. Eventually, one may construct a simplified aquatic food pyramid such that the area of each rectangle representing a particular trophic level is proportional to the estimated production of that level. But the four levels shown in this figure, as Ricker points out, are given too law a trophic value. Successive steps in a food pyramid need to be distinguished from ecological levels (groups of animals of generally similar size and habit).

In order to assess an average trophic levels one has to consider within-level predation. The effect is to reduce the net biomass produced by any ecological level. For instance, the production of lantern-fishes, which feed on zooplankton, is reduced by members of the zooplankton, such as jellyfishes and arrowworms, that feed on larval lantern-fishes. The overall transfer of production from one trophic level to the next is, of course, represented by the product KE, and Ricker estimates average values of about 10 percent for the primary consumer and stage and 15 percent for subsequent stages. Relative productivities are thus plants 100, herbivorous zooplankton 10, small pelagic fishes that feed on zooplankton about 1.6, and so forth. The more the links in a food chain, the very much less is the terminal productivity. In some oceanic communities, at least, Ryther considers there may be five links in the food chain:

Nanoplankton–micro-zooplankton–carnivorous crustacean zooplankton–larger carnivorous zooplankton–feeders on zooplankton-fish eaters.

But tuna, which feed, inter alia, on lantern-fishes, saury and euphausiids, may be put between trophic levels 5 and 6. Even so, such lengthy food chains are much less productive than those in persistent upwelling areas, as off Peru and in antarctic seas. Peruvian anchovies feed on phytoplankton as well small forms of

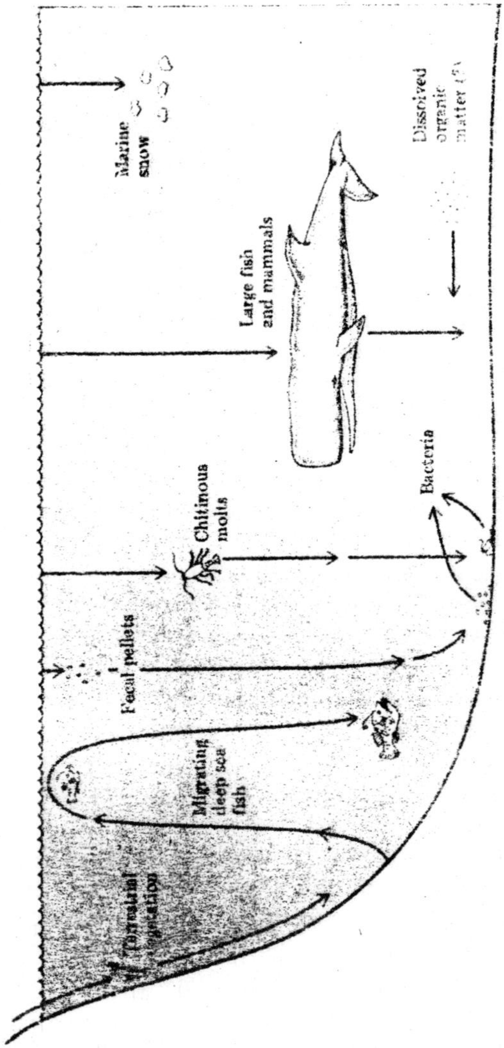

Fig. 7.4. Food sources to the deep sea.

zooplankton, while baleen whales in antarctic region feed on krill, which of all euphausiids comes closest to being a complete herbivore. Anchovies may thus be given a trophic level of about 1.5 : whales are near level 2.

Ryther, who assumed that short food chains would be the more efficient, assigned *KE* values of 20 percent to both upwelling areas, while the oceanic food chain was assessed at 10 percent. He concluded that the annual primary production of oceanic areas was generally low and gave 50 gmC/m^2/year as an average figure. The corresponding figure for the Peruvian area was estimated as 300 gmC/m^2/year. Such estimates are provisional and may well need revision. For instance, Cushing, after analysing date from the International Indian Ocean Expedition, concluded that the transfer of living material from primary to secondary trophic levels is three times as efficient in the open oceans as in upwelling areas. Perhaps there are less than five links in the oceanic food chain. If there are four rather than five, then the productivity of planktivorous fishes and piscivorous tuna will be ten-times greater than estimated. There is still much uncertainty. As Rothschild argues, one could take a number of plausible alternative values for the trophic coefficients and arrive at very different conclusions.

Mesopelagic Food Webs

The open ocean, which extends over nine-tenths of the entire oceanic area, has a total productivity that is more or less equivalent to a plant assimilation of about 15 thousand million tons of carbon per year. Investigations in the Sargasso Sea have led Riley to estimate that about 75 percent of such production is consumed by zooplankton and heterotrophic organisms in the uppermost part of the water column. Twenty percent of the total production falls to the animals and heterotrophs of mesopelagic levels, the bathypelagic zone and the bottom, Mesopelagic metabolism down to and including the level of the oxygen minimum layer accounts for most of the deep-sea consumption of organic matter, which leaves less than 5 percent to the bathypelagic and benthic organisms. Riley suggests that total amounts consumed in the bathypelagic and benthic zones are probably about equal. In the former, consumption is thinly (but unevenly) spread over a water column with a mean depth range of some 3,00 metres, whereas benthic consumption is concentrated into a narrow depth range, in on and just over the deep-sea floor.

In temperate and subarctic regions of the Pacific, which are something like three times are productive as the waters of the Sargasso Sea, Bogdanov estimated that 10 percent of the organic matter produced in the euphotic zone reaches depths below 3,000 metres, but in tropical and south temperate latitudes the corresponding figure is less than 5 percent. Clearly, there are certain outstanding question. How is organic material transported from the euphotic zone to the depths? What are the metabolic requirements of deep-sea animals and how are they met? Concerning the first question, there is the gravitational settling of particulate matter and the movements of animals, whether daily, seasonal or ontogenetic, up and down the water column.

Gravitational Transport

When particulate material in sea water is collected on a membrane filter, or in a settling cylinder, microscopic scrutiny always reveals the presence of many particles of irregular shapes and size. These particles, known collectively as detritus, consist of organic and inorganic material. Organic particles (the concentrations of which are expressed in amounts of particulate organic carbon-POC), are derived largely from metabolic waste the dissolution of pelagic of organisms, and from dissolved organic matter drawn out of solution to form organic aggregates. Coacerning the first and third sources, Sieburth under a section entitled 'Fecal Fragments', writes: 'It is obvious that much of the amorphous and flake types of suspended debris occurring throughout the water column is the debris resulting from feeding'.

The increase in suspended organic debris following phytoplankton blooms led to the concept of organic aggregates in which large reserves of DOC formed flakes and amorphous particles through a foamtower or bubble-scavenging mechanism or through film collapse. A reevaluation of bubble-salvaging indicated that it could be a laboratory artifact. It has been shown that particle formation is mainly in the <4µm range and is due to bacterial growth. This is in contrast to some 90% of the particles, which are in the 8-44µm size range.

'Not all feces or their fragments are consumed by coprophagy. Just as there is a threshold level for DOC utilization there is also a minimum threshold level for particulates below which filter-feeding organisms will not feed. This ensures a more or less steady state input of debris into the sea floor. This material as well as larger debris undergoes biodeterioration on the sea floor to form a secondary production for the development of benthic ecosystems.' The concentrations of both dissolved and particulate organic matter vary widely in the upper waters, but at depths below 200-500 metres there is relatively little change over most of the ocean. Deep DOC ranges from 05. to 0.8 mgC/litre, while the corresponding figure for particulate organic carbon is 3-10 µgC/litre. Thus, DOC is hundreds of times the more plentiful, though nearly all of it seems to be unusable by living organisms. How nutritious, then, is particulate organic matter? When, for instance, one looks in the house of an appendicularian, the gut of a copepod or the filtering-basket of a euphausiid, fragments of detritus are discernible.

Deep-sea detritus, as histochemical tests reveal, contains proteins and carbohydrates, some of which should be digestible, But laboratory tests have at best provided conflicting evidence that zooplankters, such as copepods, thrive on a detrital diet. This seems curious in view of Gordon's findings that 20 to 25 percent of deep organic detritus is hydrolysable by proteases. There is even more direct evidence from vertical profiles of the amounts of adenosine triphosphate (ATP) in the water column (Holm-Hansen and Booth, 1966; Holm-Hansen, 1970). Since ATP, the high-energy basis of metabolism, is maintained in fairly uniform concentrations in all living cells, but in soon destroyed after death, measurements of total ATP concentrations should indicate amounts of living substance. Indeed, analyses in the laboratory show that the quantity of ATP should be multiplied by 250 to give the cellular content of organic carbon (Holm-Hansen, 1970).

Investigations in the eastern Pacific off California reveal that the ATP content, which is proportional to biomass, is very high in the euphotic zone but diminishes very rapidly between depths of 100 and 200 metres, below which there is a much more gradual decline to 3,000 metres. Analyses of samples taken between 500 to 1,000 metres showed that 3 percent or less of the particulate matter was alive but above 100 metres the live fraction ranged from 17 to 79 percent. After concurring with other investigators that most of the dissolved and detrital carbon in deep water is unusable by microorganisms, Holm-Hansen suggests that a small fraction of this organic carbon turns over rapidly in support of microbial populations that may well be important as the first stage in oceanic food chains. What are the first animal stages of mesopelagic and bathypelagic food chains? At such levels the micro-zooplankton seem to be represented largely by foraminiferans and radiolarians; the remnants of which may be common in the guts of copepods, mysids, euphausiids and so forth.

Concerning the relatively large species of deep-sea zooplankton, the conclusions from Chindonova's analyses of stomach contents is that most of the 64 species examined, which included 15 out of 18 species of copepods, were carnivorous. But her results show also, together with Harding's more detailed studies, that some copepods feed at lower levels of the food chains. Harding examined deep-water copepods that were taken in closing nets fished in the Sargasso Sea and slope water off Nova Scotia. The considerable range of food items that he found in about 80 species included mineral particles and detrital remains, phytoplankton protozoans, cysts and eggs, remains of cnidarians, copepods and euphausiids. He concludes that heterotrophic protists are not only the main food of filter-feeding species of deep-sea copepods, but also form much of the diet of omnivorous and carnivorous species, In turn the heterotrophic forms '...must be supported by the dissolved and

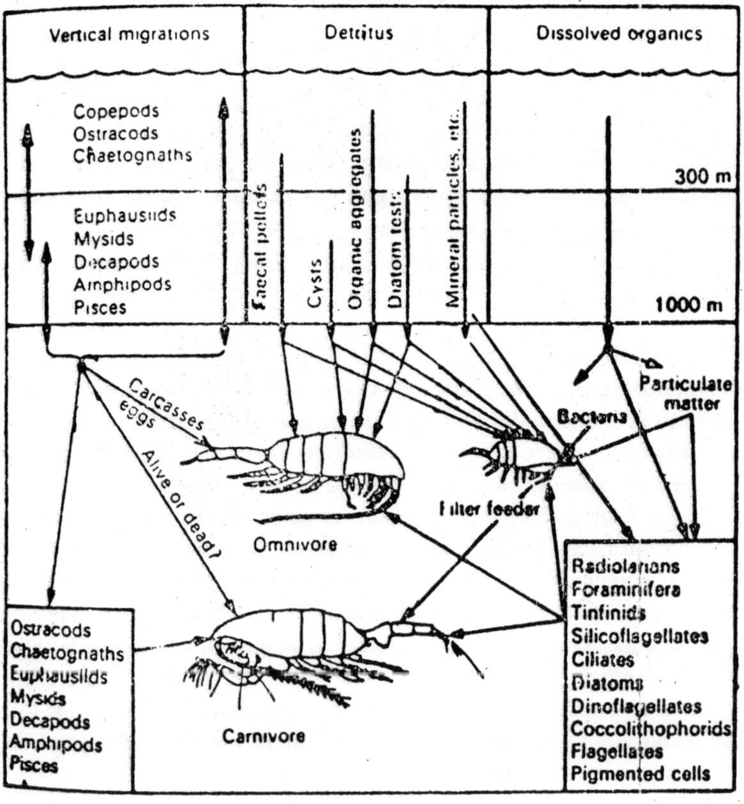

Fig. 7.5. Reconstruction of the possible and likely food sources of filter-feeding, omnivorous and carnivorous deep-sea copepods.

particulate organics, bacteria and possibly detrital remains of organisms present in deep waters'. Harding's idealized chart, which outlines the food relations of filter-feeding, omnivorous and carnivorous species of deep-sea copepods and possible energy pathways.

Further studies of this kind coupled with investigations on the quantitative distribution and metabolic requirements of the main components in midwater food chains, should resolve some of the present uncertainties. Concerning the metabolic aspect, measurements of the oxygen consumption of crustaceans and fishes from different midwater levels showed, inter alia, that species from about 1000 metres respired at a tenth of the rate of shallow-water forms. Metabolic needs are evidently modest, but how are they met?

Active Organismal Transport

Organic material is actively transported to the depths during daily, seasonal and ontogenetic movements of pelagic animals. Though the daily vertical migrations of aquatic animals was first recorded over 150 years ago, it is only now that there is enough direct evidence to show that migrants climb to more productive parts of the water column in search of food. Towards nightfall, over all main regions of the ocean, myriads of planktonic and nektonic animals make their ascent, and after spending some part of the night at higher grazing or hunting grounds, descend before daybreak to their deeper, daytime residence. In the open ocean these migrations are virtually confided to the epipelagic and mesopelagic zones. The deepest records concern daily migrations that seem to reach a depth of 1,700 metres. Every day, then, vast quantities of planktonic and nektonic forms descend from their feeding grounds, bearing means for their maintenance and growth. Vertical migrations '...certainly result in far greater mobilization of biological material than any other animal migration : using an average figure for plankton biomass of 25 gm/m^2 to a depth of 100 metres as a reasonable estimate for all oceans and assuming that diel migration adds 10 percent to this at night as a minimal figure for an average daytime residence depth 250 metres deeper, then one arrives at vertical translocation of 25 tons/km^2/day over a distance of 250 metres'. Longhurst adds that 'suth calculations are not to be taken very seriously', but his minimal figure is likely to be well exceeded in certain groups. For instance, Baker estimates that about two-thirds by weight of adolescent and adult euphausiid populations are involved in vertical migrations between 50 and 960 metres off the Canary Islands.

In taxonomic terms, Omori states that daily vertical migration seems to occur in nearly all epipelagic and mesopelagic prawns. The same is true of mesopelagic lantern-fishes. Most kinds of deep-sea migrators are cnidarians, crustaceans arrowworms, cephalopods and fishes. Relevant investigations on the vertical distribution of such migrators reveals that some species make extensive migration of several hundred metres, usually ending in the productive surface waters, while others are partial migrators. Members of a third group stay more or less at the same level throughout the day, but there are instances of reverse migrations. Such categories of behaviour may also occur in individual species. At certain times some of the adults migrate while others stay still and advanced larval stages and adolescents may be partial migrators. Concerning these and other matters, there is much to be found in the results of the *Discovery* Sond Cruise to an oceanic area off Fuerteventura, Canary Islands.

During vertical migrations animals are bound to experience changes in hydrostatic pressure. Over much of the ocean, particularly in the subtropical and tropical belts they are exposed to considerable changes in temperature and, if they live in an oxygen minimum layer, to wide ranges of oxygen tension. Moreover, in seeking biological changes, they experience others, notably the attacks of predators. But their commuting seems to be timed and regulated so as to keep the light around them at a constant intensity. There is diverse evidence that light has a controlling and dominant influence on vertical migration.

In the open ocean, daily migrations are virtually confined to the euphotic and twilight zones, and the ascent of the migrators, at least, is closely related to nightfall. Investigations of the daily migrations of deep scattering layers that are recorded by echo-sounders, and the main sound scatterers are fishes with a swim bladder, particularly lantern-fishes, supports earlier conclusions that migrators move so as to keep the ambient light field at a constant level : they are 'imprinted' on a particular isolume. During the daytime a given DSL will seek a higher level if the sky is darkened by clouds or a solar eclipse. Artificial lights will delay a dusk ascent or drive down a DSL by night. It would seem, then, that animals will only be able to shadow an isolume if their eyes are sensitive enough to perceive relatively small changes in light intensity. Mesopelagic fishes have very sensitive eyes, and so have cephalopods and decapod crustaceans. But what of migrators with much simpler eyes ? Concerning these and other aspects of light and vertical migrations, Cushing, Clarke, Kampa and Blaxter may be consulted.

The Biological Advantages of Daily Vertical Migrations

Modes of pelagic life are other than those elsewhere. The contrasts have been illuminated by Elton who wrote '...plankton has no cover structure in which predator hunts for prey, though in most other communities it seems quite likely that without such cover the community would actually collapse before long. I believe that for this reason plankton is profoundly different in its organization from terrestrial and most aquatic communities. Two features can be noted, though the whole subject merits much deeper comparative attention. For one thing, plankton has rather few species of animals living together; for another, its survival seems to necessitate very intricate 'vertical movement.' Perhaps Elton was thinking of temperate or cold seas when writing about the 'rather few species' in the zooplankton. A qualification is needed to cover the much greater diversity of plankton in tropical regions, but this simply means that the 'whole subject' needs even 'deeper comparative attention'. The biological significance of daily vertical migrations is not to be considered in isolation.

An outstanding question is this : Why have some species taken to such vertical commuting seemingly to make a living, while other species, which may be near neighbours of the migrators, habitually stay in the depths ? It would seem that one needs the concepts of ecological niche and competitive exclusion. A niche comprises the space and resources involved in a species' way of life. 'Niche breadth is simply the sum total of the variety of different resources exploited by an organismic unit'. Competitive exclusion means that no two species can occupy the very same niche. Investigations during the *Discovery* Sond Cruise, which centred on the vertical distribution and migrations of deep-sea animals in the upper 950 metres of the restricted oceanic area, close to the island of Fuerteventura are again particularly relevant.

Vertical patterns of euphausiids make a suggestive basis for comparison with other taxa, especially in view of Baker's conclusion that in principally non-migrant genera there is a decided tendency for congeneric species to occupy separate strata in the water column, whereas migrants, such as *Euphausia* spp.., have overlapping, or even coincident, vertical ranges. Recent trials in the Western North Atlantic with a multiple opening closing net have yielded samples that give strong support to Baker's generalization. Now it is evident from Roger's work in the equational Pacific that migrating

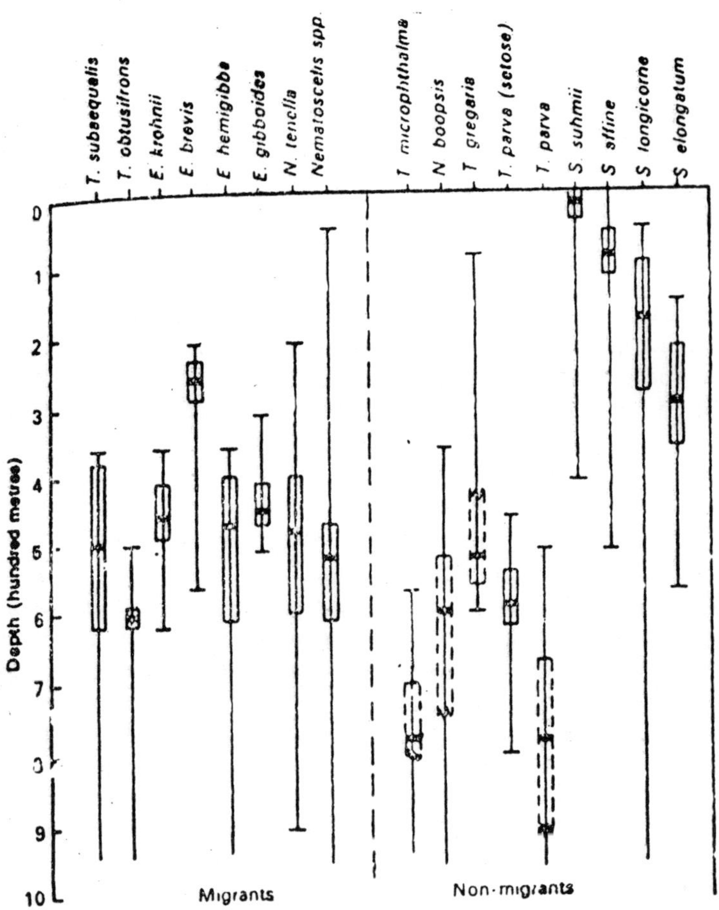

Fig. 7.6. The daytime distribution of migrant and non-migrant species of euphausiids of the Canary Islands.

euphausiids feed principally and most actively by night. If the same is true of the Canaries region, then it seems natural that migrating species should occupy overlapping or coincident vertical ranges by day. They are hardly competing for food, but merely for non-trophic living space, and evidently the different species have evolved compatible arrangements. If they feed at discrete subsurface levels by night then interspecific competition will again be minimized.

To obtain evidence of such closer-packed nocturnal segregation requires a carefully planned series of hauls, but there is some indication from Roger's investigations. For instance, in the genus *Thysanopoda*, two species, *tricuspidata* and *equalis*, migrate to levels above the thermocline, while the populations of three others, *pectinata*, *monacantha* and *orientalis*, though also found above the thermocline, are concentrated mainly between 160 and 300 metres at night. But even if *tricuspidata* and *aequalis* feed at the very same levels, they will not be competing, for the first is largely a herbivore, while the second is a carnivore. The three other species reach much the size and all are essentially carnivorous, and, as Brinton found, have nearly the same distributional ranges.

It seems reasonable that two or more related species will avoid competition if differences are evolved in their size, diet, feeding places and feeding periods. There is no evidence from Roger's study that the three above *Thysanopoda* have segregated feeding periods, but there is plenty of space, say between 100 and 300 metres, for the occupation of distinct feeding levels. Further investigation is needed. In the Canaries region non-migrant euphausiids seem to be separated by a middle mesopelagic association of migrant species into an upper group of segregated species of *Stylocheiron* and a lower (mainly lower mesopelagic) group consisting of five species belonging to three genera. Concerning the last group, Baker states that *Thysanoessa gregaria* and the typical form of *T. parva* occupy different depth ranges, with the setose form of *T. parva* lying in between.

There is also evidence from Rogers' observations that non-migrant euphausiids are less closely tied to nocturnal feeding periods than are migrants. Indeed, *Stylocheiron* spp., most of which range from epipelagic to upper mesopelagic levels, are day time feeders. Migrating and non-migrating euphausiids differ trenchantly in structural plan. The migrators, such as species of *Euphausia*, and

Thysanopoda, are generalized forms with spherical, undivided eyes and rather uniform, coarsely setose, comb-like thoracic limbs, designed to sweep or strain food from the sea. These species tend to be omnivorous, but at one end of the trophic scale there are mainly herbivorous types, at the other preferential carnivores.

All the non-migrators belong to more specialized genera that have a conspicuously elongated, anterior pair of thoracic legs, behind which is a much-reduced thoracic basket. In the most diverse genus, *Stylocheiron*, each eye is formed of a distinct upper and lower lobe, and the elongated legs are more or less chelate. These euphausiids have definite carnivorous tendencies and they are generally much less abundant than the predominantly herbivorous kinds of migrators. Migrating and non-migrating euphausiids are also generalists and specialists in other respects. Consider a subtropical or tropical species that migrates between 400 and 50 metres. Besides the possession of an adaptive organization, particularly biochemical, to tolerate considerable changes in pressure temperature and oxygen tension the species must also have the right set of physiological and behavioural attributes to spend a divided life in two quite distinct environments. It is surely a generalist. The specialist that stay at one level display no such ranging adaptability.

As regards the population structure in the water column, the generalist migrators are certainly eurytopic, the specialist stenotopic. Are there the same differences in the horizontal plane ? If the two large genera, *Euphausia* and *Stylocheiron*, are compared, and both consist predominately of epipelagic to upper mesopelagic species, it emerges that species of the former cover a much wider area. *Euphausia* is represented in the substractic, antarctic and subarctic zones and in the subtropical and tropical belts while *Stylocheiron* is essentially a warm-water genus (Brinton, 1962). Thus, evolutionary potentiality in the genus *Euphausia* has been such that species have emerged that are not only adapted to the wide range of physical and biological conditions associated with diel vertical migrations in the warm ocean, but also, inter alia, to the widely ranging seasonal conditions, especially biological, in colder regions.

One is reminded of the recent work of Jackson, (who finds that the infaunal bivalves adapted to the rigours of life in depths of 1 metre or less have distinctly wider geographic ranges than other infaunal species restricted to depths greater than 1 metre. Again, the set of species with wider environmental tolerances has the greater geographic representation. Sergestid prawns, some of which prey on euphausiids, are active migrators. Off Fuerteventura, Foxton found that species of the subgenus *Sergeste*, which are semitransparent and bear internal, ventral light organs, live at levels above 750 metres, whereas *Sergia* spp., uniformly red and with external light organs, live below this level.

Here, in a daytime descent from 400 metres and considering only the migrators, there are three pairs of species that each live together in concentrations between the same levels of the water column; *Sergestes sargassi* and *S. armatus S. vigilax* and *S. pectinatus* and *Sergia robustus* and *S. splendens.* There is also *segestes corniculum.* Partly or entirely below the last pair, there are two non-migrating bathypelagic species, *Sergia japonicus* and *S. tenuiremis.* For investigations on both the vertical pattern and food of sergestids, we turn to the Ocean Acre project off Bermuda. During the day there is again a step-down occupation of the water column with groups of two or more migrant species concentrated between the same levels : *Sergestes cornutus* and *S. sargassi*: *S. pectinatus, S. vigilax* and *S. armatus, S. atlanticus S. corniculum* and *S. grandis* and *Sergia robustus* and *S. splendens.* Again there are the same bathypelagic species, *Sergia japonicus* and *S. tenuiremis*, but in the Bermuda area they appear to be partial migrators. During both day and night the only species that occurred together were *Sergestes sargassi, S. pectinatus* and *S. vigilax*, but dissimilar features of their third maxillipeds, a key pair of limbs involved in feeding, led Donaldson to suggest that they have different diets.

Donaldson found that most species took a variety of food organisms, but three species contained euphausiids more often than other kinds of prey. Fish were most frequently found in *Sergia robustus* and *grandis.* The bathypelagic species, *Sergia japonicus*, which has a flaccid body with reduced musculature, appears to be a detritus-feeder. In most species food items were found most often at night, but no species was exclusively a nocturnal feeder. On the other side of the Atlantic, however, Foxton and Roe who analysed the food of six sergestid and three oplophorid prawns, concluded that intensive night-time feeding seemed to be the rule in all nine species. Once more, partitioning of resources and the water column by a set of migrating species seems to be such as to reduce interspecific competition.

During the daytime in particular, groups of two or more species may coexist in relatively narrow strata, which is quite unlike the mutually exclusive stratification in the non-migrating euphausiid, *Stylocheiron.* How far do copepods show the same trends ? They are the prey of innumerable marine animals from cnidarians to whales (including) euphausiids and sergestids) and at any one point in the warmer parts of the ocean, many species occupy the water column from top to bottom. In the Fuerteventura area 212 species of calanoid copepods were identified in hauls made between 960 and 40 metres. The bulk of the population, which consisted largely of immature stages, was centred at 500 metres by both day and night. Of the adult forms, some were conspicuous migrators but most species showed restricted migratory tendencies or were non-migrators. Considering first the non-migrators, it is evident that related species do not necessarily follow the discretely stratified pattern. For instance, three pairs of species : *Calanus belgolandicus* and *C. carinatus*; *Rhincalanus nasutus* and *R. cornutus* and *Spinocalanus magnus* and *S. spinosus* were concentrated between much the same depth ranges throughout the entire day.

After the name of each species in parentheses is the length range of the females and it will be seen that in each pair one species is decidedly the larger. The significance of size differences in reducing competition between coexisting, congeneric species of animals has been well demonstrated by Hutchinson who found that the ratios of mouthpart sizes in relevant species of insects, birds and mammals ranged from 1.2 to 1.4. As Wilson observes, larger predators are able to utilize food sizes that are not available to smaller predators, while the reverse is much less true. Within a species competition for food may be reduced by sexual differences in size but in most oceanic copepods the males are slightly the smaller sex. On the other hand, the adult males of some species take no food. There is every indication, then, that Hutchisons's rule also applies to certain coexisting, congeneric species of oceanic copepods. Species of *Undeuchaeta* and *Pleuromamma* are among the conspicuous migrators.

By day *Undeuchaeta major* and *U. plumosa* were concentrated between 450 and 900 metres by night the migrators had reached levels above 250 metres. But if most of their food is taken at night there need be little competition, for once more, the size difference between the females (in parenthesis) is considerable. Two common species of *Pleuromamma, xiphias* and *abdominalis* had overlapping distributions by day, the first centred at 570 metres, the second at 400 metres. By night most of the extensive migrators were clustered around levels of 250 and 100 metres, but yet again, size differences are such as to reduce competition. Moreover, in the uppermost waters Roe has evidence to suggest that *xiphias* was most abundant between 50 and 25 metres, *abdominalis* between 10 and 0 metres. Lastly, three other species, *borealis pisekei gracilis*, are centred at 350, 300 and 200 metres, respectively, during the day; by night the first is concentrated at 250 and 100 metres, the other two at 50 metres.

Nocturnal competition between *pisekei* and *gracilis* is possible. They are of much the same size and all three species, as Roe observes, are closely related. He adds also that samples from the uppermost 50 metres showed that *borealis* was scarce and *pisekei* more abundant than *gracilis*. Further investigation should be of interest. The evolutionary interplay of mesopelagic life seems to have been such the potential competitors in the zooplankton and micronekton have come to be more or less segregated through mutually exclusive combinations of spatial, temporal and biological attributes. The same is true of the productive upper waters, as may be seen in Zalkina's (1970) paper on the vertical distribution of cyclopoid copepods in the equatorial western Pacific. During a 24-hour station a series of hauls was made between the surface and 300 metres. The catches showed that the species of Oithonidae, which are non-migrators and concentrated between 75 and 150 metres, have little overlap in the water column with those of the families Oncaeidae and Corycaeidae.

It is interesting also to compare the changing vertical dispositions of the two dominant species, *Oithona plumifera* and *Oncaea media*. Again, there is a little overlap, except from 10.00 to 18.00 hours, when the second species is keeping its daytime station. Zalkina found that both migrating and non-migrating species tended to maintain a step-like distribution. Some species differed in the times of their ascent and duration of stay at higher feeding levels, and Zalkina concluded, 'This peculiarity of disposition tends to lessen the intensity of food competition between species with similar nutrition habits. 'She also quotes her colleagues, particularly M.E. Vinogradov, who also found that dominant species of oceanic zooplankton with similar food requirements occupy discrete strata in the water column, and that such segregation is particularly marked in regions of low productivity. The Californian sardine is an active fish, and during its virtually restless life, respiration consumes most (82.99 percent) of the calorific content of its hard-won food in the zooplankton.

Reproductive processes take a mere 1 percent of assimilated calories. The lives of mesopelagic fishes are not so strenuous, but whether they migrate or not, they are highly organized predators, needing the means to run elaborate tissue systems, particularly the substantial muscles of their jaws and motor complex. How do they share available resources ? Most kinds of mesopelagic fishes are lantern-fishes, alepisauroids or stomiatoids. Of all the twelve families that form these groups, daily vertical migrations are by far the most prevalent in the lantern-fishes. Indeed, except for some of the deepest-living kinds, such as certain *Lampadena* species and those of the genus *Taaningichthys*, well-planned investigations reveal that lantern-fishes are inveterate migrators. Judged by their biomass, ubiquity and species diversity, they are also the most successful family of midwater fishes. In relative length and other features of their jaws, which bear rows of small conical teeth, there are three main groups; small-jawed, medium-jawed and long-jawed. The genus *Diaphus* is unusual in having representatives in all three groups (Foxton, 1972). One species, *Diaphus taaningi*, a rather small lantern-fish with medium-sized jaws, is known from the Gulf of Mexico, the Caribbean Sea and off West Africa.

In the Caribbean it lives in the Cariaco Basin, a depression in the continental shelf off Venezuela with a shallow northern still that restricts the horizontal exchange of water between Basin and Sea. Due to this restricted circulation and other factors the Cariaco Basin is anoxic below a depth of about 300 metres. But *D. taaningi* lives above the anoxic water. Baird and his colleagues found that it migrates from a daytime depth of about 250 metres to appear in the upper 50 metres at night. Nearly all of their fish ranged from 38 to 45 mm in standard length, and all were feeding largely on herbivorous zooplankters, particularly copepods and larvaceans, which formed about 70 percent of the total prey biomass.

Analyses suggested that most of the prey was taken in the early evening, leaving the rest of the night for digestion. Fish taken before the evening ascent were essentially empty. In their discussion the authors argue that *Diaphus taaningi* and certain other midwater fishes, such as the hatchet-fish, *Polyipnus*, are bound to near-neritic waters where there is an abundance of small zooplankton. 'As prey density and distribution at a given time are thought to determine the amount of food a predator can obtain as well as the cost to the species in energy of pursuit, it would appear that small prey in high density in the night time depth zone may well be a niche requirement for *D. taaningi.*' Two other lantern-fishes, *Stenobrachius lcucopsarus* and *Triphoturus mexicanus,* which are long-jawed and larger than *D. taaningi* were the dominant midwater fishes taken during a cruise off California. Collard, who found that both preyed mainly on small and medium-sized crustaceans, particularly copepods and euphausiids, concluded that diets were diverse, with no sign of selection. But some mesopelagic fishes feed selectively, as Merrett and Roe (1974) found after analyses of fish caught by *Discovery* in a series of eleven hauls over a 24 hours period between 230 and 266 metres in the eastern North Atlantic.

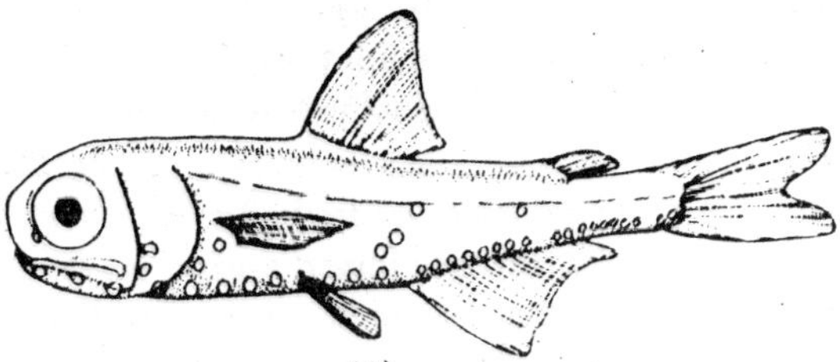

(A)

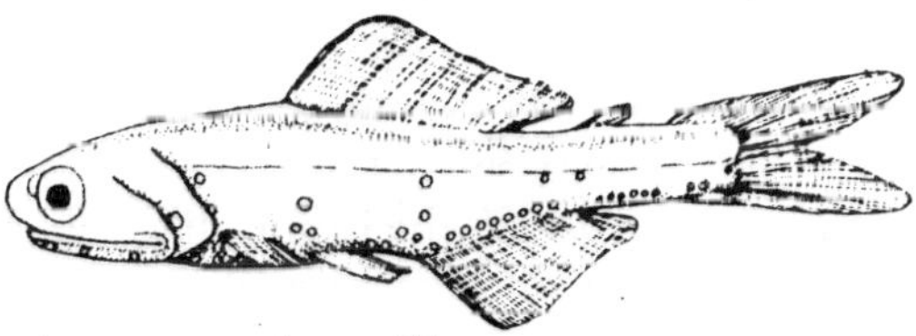

(B)

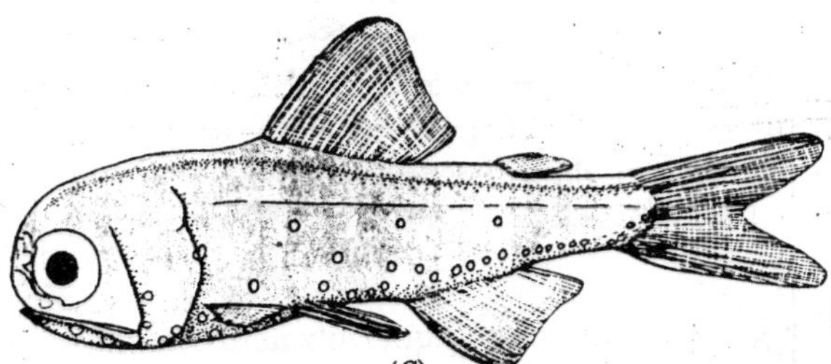

(C)

Fig. 7.7. Large eyes in three mesopelagic fishes of the family Myctophidae. (A) Myctophum punctatum. (B) Lampanyctus elongatus. (C) Diaphus metopoclampus.

By comparing the composition of the zooplankton (in the net) with fish prey, the authors were able to determine whether or not food items had been selected by some of the more abundant fishes. They looked also for periodicities in feeding. The small stomiatoid *Valenciennellus tripunctulatus*, virtually a non-migrator, fed mainly by day on selected calanoid copepods. Two partial migrators, the hatchet-fishes, *Argyropelecus hemigymnus* and A. *aculeatus*, which seemed to feed mainly at dusk, differed in diet. The first fed randomly on mixed copepods and ostracods, the second selectivity on ostracods.

There were parallel results from two fully migrant lantern-fishes, *Lobianchia dofleini* and *Lampanyctus cuprarius*. Both fed at night, the first randomly on copepods, the second on selected amphipods, particularly the males of *Phrosina semilunata*. The viper-fish *Chaulidous danae*, an avid predator, seemed to feed at any time and randomly. Merrett and Roe concluded that vertical migrations and feeding habits are closely connected. Whether by the migration of prey species away from levels occupied by their predators or by the upward migration of the predators to the headquarters of their prey. At the same station another series of nets was fished down to 2,000 metres, and from these catches Badcock and Merrett have presented an overall analysis of vertical patterns in the midwater fish fauna. Considering first the lantern-fishes, their results complement those of Badcock from the Fuerteventura area off the Canary Islands. All species, or at least all well represented species, were diel migrators. By day the vertical ranges of silvery species largely overlap or are coincident between levels of 300 and 700 metres, while dark species have similar range correspondences below 700 metres.

Silvery lantern-fishes are transparent species of sergestid prawns have much the same depth range : so do dark lantern-fishes and red sergestid prawns. Indeed, lantern-fishes and migrating species of sergestid prawns and euphausiids coexist at similar or widely overlapping depth ranges by day. They also tend to take most of their food by night. There is also evidence of dietary differences amongst them. Over most regions of the ocean between the subarctic and subantarctic zones the most numerous non-migrators at mesopelagic levels are the more or less transparent species of *Cyclothone*, stomiatoid fishes known as bristle-mouths. Dark *Cyclothone* live deeper and like other bathypelagic animals, which are considered later, are also non-migrators.

The transparency of mesopelagic bristle-mouths is interrupted by black markings, especially at internal levels, but in *Cyclothone pseudopallida* there is also an integumentary dusting of small melanophores over much of the body. When adult they have a gas-filled swimbladder, which regresses and becomes fat-invested in their dark relatives. Further differences between the two groups are considered elsewhere. At mesopelagic levels in the Canary Islands region there are two 'transparent' *Cyclothone, braueri* and *pseudopallida*. There is also a third species *pallida*, which ranges from lower mesopelagic to bathypelagic reaches. The first is concentrated between 400 and 600 metres, the second between 500 and 800 metres and the third between 600 and 800 metres. There is thus relatively little overlap between *braueri* and the other two species and *braueri* is not only the dominant species but also the smallest (maximum standard length 38 mm v. 45 mm and 63 mm).

Competition between *braueri* and the deeper-living pair is evidently slight and is reduced also between mature males and females of *braueri*, to consider only the dominant species. As in other *Cyclothone*, the males are not only the smaller sex but nearly

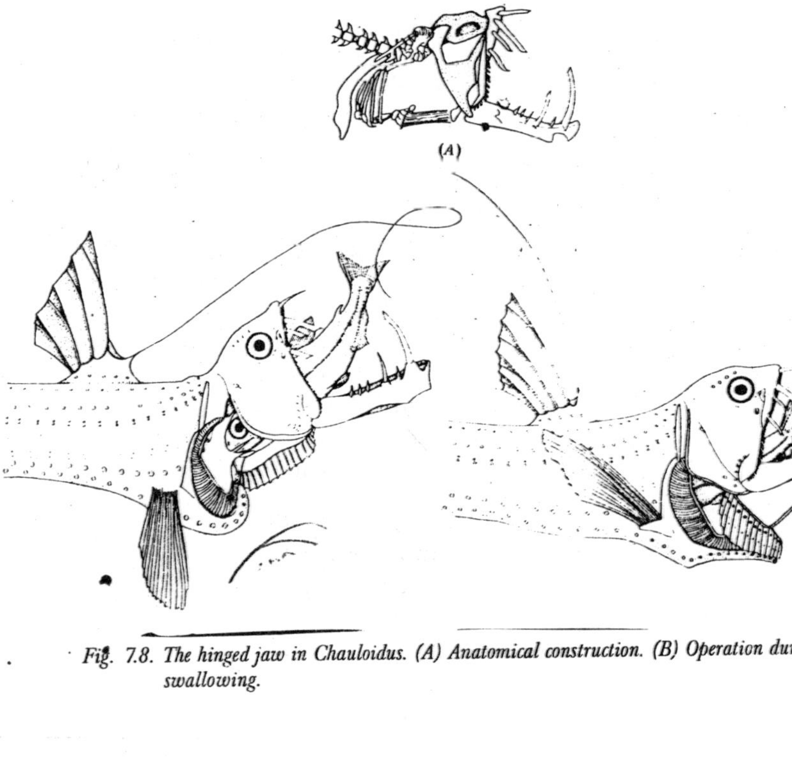

Fig. 7.8. The hinged jaw in Chauloidus. (A) Anatomical construction. (B) Operation during swallowing.

all live at higher levels than the maturing to ripe females. Apart from reducing intraspecific competition, the segregation of the sexes may well facilitate eventual sexual congress. Competition between mature individuals of *pseudopallida* and *pallida* is likely to be modulated by their difference in size, but other information is needed, particularly on their diet and trophic periodicity. The concentrations of these two species also overlap with the upper part of the depth range of *Cyclothone microdon* but this is made largely of young individuals which are not likely to compete much with mature fishes of the other two species.

Lastly it will be recalled that Badcock and Merrett's paper is based on sampling at 30°N, 23°W, about 600 miles west of Fuerteventura, Canary Islands. At this station *Cyclothone microdon* is a dominant species, but off Fuerteventura it was not recorded. Faunal composition in the water column changes from place to place but there is still the interplay of non-migrating species to reduce competition. Off Fuerteventura the relations *between C. braueri, pseudopallida* and *pallida* are rather like those at the westerly station, though there is less overlap between the second and third species, and *microdon* is replaced as it were, by two other dark-coloured bathypelagic species, *C. acclinidens* and *livida*. Seven of the nine dominant species of lantern-fishes in the Bermuda area are small or diminutive (Gibbs *et al.*, 1971), about the same size as the mesopelagic species of *Cyclothone*. Like *Diaphus taaningi* these lantern-fishes must be restricted by their relatively small gapes to small prey organisms.

If any one of the seven is compared to a *Cyclothone* of the same length, the latter will be seen to have a considerably wider gape, for the jaws are about four-fifths the length of the head, which is about a fifth to a quarter of the fishes standard length. The *Cyclothone* should be able to take a wider size range of prey items than can a diminutive lantern-fish. But the latter migrates up to epipelagic levels or just below, where there is likely to be a greater abundance of small prey, particularly of herbivorous copepods, both young and adult. To make a living at deep-sea levels, where, with increasing depth, there is more or less exponential decline in the biomass of suitable food, it would seem that a non-migratory fish should be adapted to secure the widest possible size range of food organisms.

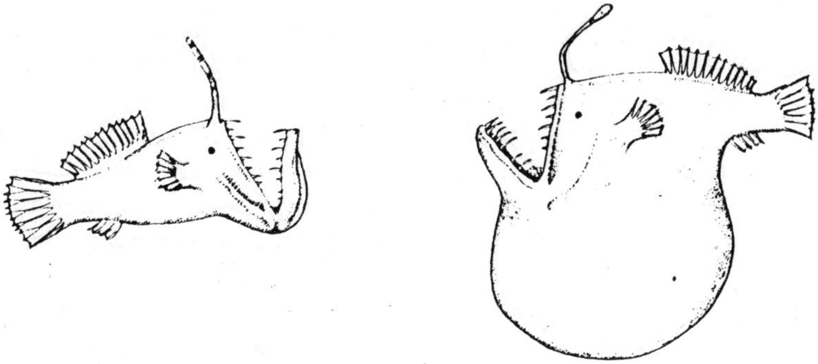

Fig. 7.9. The deep water angler fish, Melanocetus johnsoni, before and after swallowing a larger fish.

Indeed, investigations off southern California showed that *Cyclothone acclinidens*, which lives below the transparent species, *C. signata*, at depths greater than 400 metres, takes a wider size range of prey from copepods to arrowworms and amphipods. This is understandable, considering that the deeper-living species bears the larger jaws, due both to its greater size (maximum standard length 50 mm v. 33 mm) and to its relatively larger head. Similarly, Ebeling and Cailliet found that the deepest living and non-migratory fishes called big scales not only have larger jaws than shallower-living species, but also develop a gill raker system fine enough to retain and concentrate food items as small as 1 mm or so. The larger a fish, the less energy it needs per unit of body weight. Even so, a large fish cannot exist in a food-poor environment : there are no deep-sea tunas and billfishes. Some of the largest mesopelagic fishes are alepisauroids, but unlike their relatives the lantern fishes, they are nearly all non-migratory.

The lancet-fish *Alepisaurus ferox* grows to about 2 metres, but it is not a heavy fish. A five-foot individual weighed about 4 pounds, which is not only indicative of its long compressed body but especially of its lightly ossified skeleton and sparely developed motor muscles. It is above all due to the reduction of skeletal and muscular systems that deep-sea midwater fishes nearly attain neutral buoyancy. Thus, as well as being economically made, they need little energy to keep to a preferred level in the water column. The barracudinas, which range in length from a few to over 50 cm, have both the look and economical build of their relatives, the lancet-fish : so have other alepisauroids. These fishes have also the jaws, swallow and maw to take very large meals. An individual of one species, *Anotopterus pharao*, which was swallowed by a whale captured in the Ross Sea, Antarctica, and measured nearly 75 cm, contained two barracudinas of 18 and 27 cm in length. The barracudinas were bloated with krill, *Euphausia superba* (Marshall, 1955). More recently, Wassersug and Johnson (1976) have shown that alepisauroids have a sac-like stomach that is both large and greatly distensible.

The intestine is simple and straight. Species of all six families swallow intact prey that are at least one-third to one-quarter of their length. As the two authors say, in energetic terms midwater predators in the deep sea must not be much larger than their prey. But given the above and other adaptations alepisauroids can afford, at it were, to be non-migrators. But there are non-migratory, tiny-mouthed mesopelagic fishes. Such argentinoids have large upturned tubular eyes with parallel optical axes, which, together may well provide extra sensitivity. Even more important, such eyes presumably afford much-needed accuracy in pin-pointing small prey that is to be sieved in tiny jaws.

Hatchet-fishes of the genus *Argyropelecus*, which at most are partial migrators, have similar eyes but larger jaws. Even so, accurate prey taking is still needed, particularly when the diet is selected, as in *A. aculeatus*. Where prey of a suitable size may be sparse, chances of nourishment should not be missed. And there is a further aspect. The hatchet-fish *Argyropelecus affinis* and the alepisauroid *Scopelarchus analis*, a voracious fish, also with tubular eyes, have yellow lenses. The function of such lenses may well be to enable their owners to 'see through' the bioluminescent camouflage of various midwater animals. Meals may be seen that might be missed by predators without special filters. More evidence could be adduced to support the idea that diverse mesopelagic fishes have each evolved a unique and special organization to enable them to stay in the depths and make a living after their metamorphosis to adult form. The same is true of non-migratory euphausiids.

Migrators are generalists, adaptable to wide ranges of physical and biological conditions not only vertically but also in the horizontal plane. Is this also true of fishes? For instance, one can compare the distribution of the migratory lantern-fishes with their non-migratory relatives, the alepisauroids. To take but one criterion of extensive distribution, about 15 species of lantern-fishes are known from the antarctic zone but only three or four alepisauroids. Daily vertical migrations, evidently in search of food are thus ingrained in many kinds of epipelagic and mesopelagic animals.

One advantage is that during their resting, non-feeding periods, potentially competing species can reside together between much the same depth levels. Indeed, the entire nexus of vertical migrations would seem to be fundamental to species packing and community structure in the water column. Both these aspects are well illustrated by fishes, particularly as they are evidently the outstanding components of the deep scattering layers recorded on ships' echosounders (see Backus and Crad dock, 1977, and Hersey and Backus, 1962). The prominence of fishes as acoustic scatterers is due largely to their gas-filled swimbladder, which is present in most lantern-fishes and other mesopelagic forms, notably in the smaller stomiatoid fishes (Marshall, 1951, 1960 and 1970). Some scattering layers undergo daily vertical migrations, but others are more or less stationary.

In the present context a ranging survey of DSLs in the North Atlantic by Haigh (1970) will be considered. Records of a 10 kHz echo-sounder from 1963 to 1967 and between latitudes 10° and 68°N revealed five mesopelagic layers during the day. The first three (ranges of maximum mid-layer depths : A 135 – 175 fms, B 140 – 245 fms, C 215 – 280 fms) were purely migratory and found in latitudes south of 49°N. Layer D, which was partially migratory and attained a maximum depth of about 330 fms, was the most widespread, being recorded at all latitudes. Layer E. was mainly migratory and recorded only in restricted areas in latitudes south of Iceland. Thus, north of 49°N there was one main layer D. joined locally by layer E. South of this latitude there were up to five layers and over wide areas three occurred together.

The conjecture is that the number of DSLs is a reflection of species diversity, which is much greater under the armer waters south of ca. 40°N than in the colder waters to the north. Where there are more species there is need for more statified species packing in the water column. Indeed, considering, for instance, that there are over 200 species of mid-water fishes off Bermula it is remarkable that there are so few deep scattering layers. That there are so few would seem to be related to the phenomenon of daily vertical migrations, whereby sets of potentially competing species can live together between the same depth ranges when they are resting by day. During the might, all but the non-migratory and partially migratory layers are in the upper 150 metres, where broad and prominent sound scattering is recorded. The biomass and number of species in a DSL may be greater than those outside but do the sound scatterers and associated species form communities? Off Santa Barbara, southern California, factor analyses of the catches of deep midwater animals revealed a complex of four resident communities, which interact with a number of transitory, presumably less stable, groups. On a larger scale there are investigations of Backus and his colleagues in the Atlantic. Daylight hauls were generally made at the depth of sound scattering maxima, while might hauls were set to sample concentrations of migrating fishes at depths above 300 metres.

Though largely concerned to sample layers of strong sound scattering, the investigators are readily able to refer the distributions of the mesopelagic fishes in their catches to six or more pelagic regions, each with a '. . . more or less unique fish fauna with its characteristic assemblage of species in characteristic proportion, its characteristic diversity and so on'. This definition could also be applied to the communities that presumably contribute largely to the individual characteristics of a pelagic region. In turn, the more or less stable structure of a community is integrated by the many relationships between its constituent species (particularly predator-prey interactions and interspecific competition) and the environment.

With such community criteria in mind, it will be interesting to see how far they apply to the animals in sonic scattering layers. There is much to do. As Kinzer says, we know little about the predator-prey interactions of organisms within the diurnal range of such layers. If daily vertical migrations are feeding sorties, the metabolic benefits must exceed the energetic costs of ascent and descent. No doubt the benefits will vary from day to day, but even so, the energy needed for vertical migrations by small crustaceans is but a few percent of their daily metabolic rate. For planktonic crustaceans ranging from 0.5 to 30 mm in length, the amount of energy expended during vertical migrations was calculated (by estimating the work required to over come hydrodynamic resistance and lift their net body weight) to be no more than 20 - 40 percent of their resting metabolism. Comparable figures for small fishes may well be of the same order. The adaptive value of daily vertical migrations in energetic terms has been considered by McLaren.

In a study of migrating zooplankton McLaren followed the implications of a mathematical model based on von Bertalanffy's growth equations. The model predicts that a migration from cool to warmer waters can be advantageous : an animal that is mainly active when feeding at upper levels and then descends to rest in cooler waters, will use its food to good effect. For instance, growth and fecundity should be enhanced, for during its resting periods an animal should eventually grow larger than it would in warmer surroundings. Though subsequent tests of this hypothesis have not been encouraging, it should not yet be abandoned. Besides the gain in food, the lessening of competition and the partitioning of resources, there are other likely adaptive aspects of vertical migrations.

For planktonic animals, at least, Hardy, after pointing out that during their migrations currents may carry them considerable distances in the horizontal plane, suggests that, '... the development of a habit of regular periodic vertical migration has been evolved to enable the animal continually to sample new environments–new feeding grounds – and so to have some power of choice, of varying its behaviour within limits'. Following this line of thought, Isaacs, Tout and Wick have shown how animals may be transported into areas of high standing crops of phytoplankton. Besides gaining food, vertical migrators may escape predators. For instance, in the tropical Pacific yellow fin tuna and albacore evidently feed during the day, when lantern-fishes and other likely prey are residing at lower levels (Roger and Grandperrin. 1976).

Though one cannot imagine lantern-fishes with their sensitive nocturnal eyes in the surface waters by day, it seems also that they avoid lancet-fishes, voracious mesopelagic predators. In both the North Atlantic and the southeastern Pacific the fish prey of *Alepisaurus* consists largely of hatchet fishes, barracudinas and others of their kind (Haedrich and Nielsen, 1966). All such prey are non-migrators; lantern-fishes, nearly all of which migrate, are rarely taken. Clarke also observes that studies of the feeding habits of tuna, dolphin-fishes (*Coryphaena*) and *Alepisaurus*, indicate that they consume very few vertically migrating forms such as lantern-fishes and gonostomatid fishes. Thinking of epipelagic predators, particularly yellow fin tuna and albacore, Roger and Grandperrin argue that if such predators do not feed at night, they cannot retrieve the energy won by the migrating mesopelagic fauna. They see the whole process as an 'energy valve' permitting the downward transfer of energy but reducing the upward one. They conclude that if this process exists in other parts of the food chain', ...then one can suggest that the mechanism is fundamental, as it explains how the deeper layers of the ocean are able to support a fairly large biomass'. Their ideas seem well worth pursuing.

Lastly, ontogenetic and seasonal migrations also lead to transfers of organic matter between the euphotic zone and mid water levels. Concerning the former, we shall see that in many kinds of mesopelagic and bathypelagic animals the spawning place and buoyant properties of the eggs are such that the larval life history takes place in the productive surface waters. During and after metamorphosis there is a descent to the adult-living space. There must be heavy mortality as the eggs and early larvae ascend to their euphotic nursery ground and during their stay in this layer. The descending metamorphosis stages will also fall to predators, but the young are now large and active enough to avoid some of their enemies. Seasonal movements of zooplankton between epipelagic and lower levels seem to be largely confined to temperate and cold waters and upwelling regions.

Conditions in antarctic waters are considered by Foxton, whose investigations showed that the bulk of the standing crop of zooplankton was concentrated in the per 100 metres during the summer months of November to March. In the underlying waters there were also seasonal variations such that maximum quantities were reached from July to September, when the standing crop in the upper 100 metres was at a minimum. But over the whole water column from 0 to 1,000 metres there were no significant monthly variations in the mean total volume of plankton. Foxton continues; 'This perhaps unexpected result can be explained by what is known of the life histories of some of the Antarctic zooplankton. It has been shown that one of the most marked features of the plankton cycle in Antarctic waters is that some of the commoner species, including the copepods *Rhincalanus gigas* and *Calanoides acutus*, and the chaetognaths *Eukrohnia hamata* and *Sagitta gazellae* migrate seasonally so that the bulk of the adult population during the summer is concentrated in the surface layer and during the winter in the waters of the warm deep current'. Indeed, Mackiatosh proposed that such migrations are of biological significance in that species populate the northerly flowing surface water in the summer but return south in the counter-flowing, warm deep current in the winter, thereby staying within their proper geographic range.

Bathypelagic Food Webs

The bathypelagic zone, the largest environment on earth, is cold and dark and the most deserted life zone in the ocean, both in numbers of organisms and of species. Apart from their ontogenic migrations, the deep set ways of bathypelagic animals would seem to give them little or no direct contact with the more productive waters above their habitat. There may well be overlap between populations of lower mesopelagic and upper bathypelagic animals. Thus, dark-coloured lantern-fishes, most of which live at lower mesopelagic levels and undertake daily vertical migrations, are sometimes eaten by female angler-fishes. But virtually nothing is known of the exchange of organic material in the overlapping zones. Certainly the zooplankton biomass near 1,000 metres is considerably greater than that at lower levels, but the main factor may simply be depth of residence; those highest in the bathypelagic water column are the first recipients of organic crumbs from above and thus have the means of highest productivity, Copepods, which must be the main metazoan users of organic particles, are well represented at bathypelagic levels.

At all depths copepods, including their copepodid stages, make up the bulk of the zooplankton. Indeed, off Bermuda, Deevey and Brooks found that the relative proportions of copepods in the total zooplankton catch increased with depth in the upper waters to 91 percent between 1,500 and 2,000 metres. The decrease in numbers and quantities of zooplankton is greatest between the upper waters and a depth of about 500 metres the vertical gradient for volumes or numbers of zooplankton falls exponentially, the rate of decline being greater between 503 and 2,003 metres than that below this depth range. For the biomass of total zooplankton, as measured by displacement volumes, Deevey and Brooks found that the slope of the gradient between 500 and 2,000 metres was – 5.66 × 104 log units per metre, which they compare with Vinogradov's figure of – 3 to – 2 x 104 log units per metre for the gradient between 2,000 and 4,000 metres in the Indian Ocean. There is also a decline with depth in the numbers of species. For copepods; Grice and Hulsemann recorded 153 bathypelagic species in the western Indian Ocean, of which 122 occurred between 1,000 and 2,000 metres, 73 between 2,000 and 3,000 metres and 13 between 3,000 and 4,000 metres. Of the 326 species they identified from catches made between the surface and 2,000 metres off Bermuda, Deevey and Brooks recorded 128 species between 0 and 500 metres, 204 between 500 and 1,000 metres, 172 between 1,000 and 1,500 metres and 119 between 1,500 and 2,000 metres.

At upper levels and in an area of high productivity it may well be, as Harding (1974) found, that protists are an important food source for bathypelagic copepods. But in the poorly productive Sargasso Sea at deeper levels his analyses show that the most frequent gut contents of copepods were mineral particles, detrital remains and detrital balls. Deevey and Brooks conclude that phytoplankton production in the Sargasso Sea is insufficient to support resident populations of filter-feeding copepods. Throughout the water column, they suggest that most copepods are omnivorous opportunists, consuming whatever particulate material, microorganisms and nauplius larvae are available. They contrast their findings with those of Vinogradov from the Kurile-Kamchatka region, where carnivorous species increase in importance between 1,500 and 3,000 metres while omnivores prevail at lower levels. But, as they say, such trophic stratification is probably typical of regions of high primary productivity.

Clearly, there is much we do not know of bathypelagic food pyramids, but recent investigations are suggestive. Light scattering measurements of the standing crops of suspended particles at bathypelagic levels in the Atlantic Ocean reveal a quantitative distribution that matches the pattern of surface productivity which is also reflected in the standing crops of bathypelagic and benthic organisms. Even more significant, we shall see later that the faecal products of copepods and other zooplankters, especially from the epipelagic zone, are a major source of food for benthic, deep-sea animals. Many such waste packages must be intercepted and used by bathypelagic zooplankton. The greater the surface productivity the more will be the 'snowfall' of organic particles and the 'hail' of faecal materials, and both together, particularly the latter, may well go far to determine levels of productivity in bathypelagic and benthic environments. At mesopelagic levels, as we-have seen, there is also the powerful influence of vertical migrations.

The copepods, which are so predominant numerically at bathypelagic levels, must go far to nourish resident populations of arrow-worms, euphausiids prawns and diverse fishes. Apart from the small *Cyclothone pygmaea* in the Mediterranean and *C. livida* off Western Africa, black species of *Cyclothone* range over the three oceans from lower mesopelagic to bathypelagic levels.

The open ocean species are not only larger than the transparent mesopelagic species, but they have a relatively larger head and mouth. Their prey ranges widely in size from other *Cyclothone* through euphausiids, amphipods and arrow worms to octracods and copepods, As stressed already, where food is scarce it is fitting to have the means for securing such a ranging regimen. Thinking particularly of this desideratum, we turn to the deep-sea angler-fishes, which are the most diverse group of bathypelagic fishes: indeed, the 100-odd species make up at least two-thirds of the bathypelagic fish fauna.

Between adolescence and when they are ready to seek their partners the free-living dwarf males have relatively small jaws and a more or less fusiform body. Unlike the females, they have no angling system. In some families the males, which eventually become parasitic on the females, are not known to take food after metamorphosis. Males of other families have jaws that are not only suitable for taking prey but also, presumably, for hanging on to the skin of a female during the breeding season. Such males grow after metamorphosis and those of *Melanocetus* at least, feed on zooplankton. The larger females, which is most species tend to a globular shape, bear large jaws armed with recurved, depressible teeth. The length of the jaws is usually more than half the length of the head. Except in *Neoceratias*, all females bear a rod and attached bait containing in nearly all species a luminous gland. The illicium projects from the head between or just before the eyes and is articulated to a horizontal basal bone set in a groove on the cranium. There are muscles from the basal bone to the base of the illicium such that the latter can be moved between an upright and a horizontal position.

Moreover, in some angler-fishes the basal bone is so long and movable that the rod and bait can be extended well in front of the mouth and then retracted Williams rightly urges critical circumspection before deciding that features have an adaptive significance and he writes, 'A frequently helpful but not infallible rule is to recognize adaptation in organic systems that show a clear analogy with human implements. The angling system of ceratioids is surely an outstanding example.' Female angler-fishes feed largely on crustaceans and fishes. They also take arrowworms and squid.

As in *Cyclothone* an individual may contain organisms of a wide size range. Thus, in the stomach of an adolescent *Himantolophus* there were small copepods, two amphipods, five euphausiids, four *Cyclothone* one hatchet-fish one lantern-fish, one melamphipods fish one 'sudid' fish and one squid beak. Moreover, the tissues of the stomach and abdominal wall are distensible enough (to accommodate very large meals which may consist) largely or entirely of one prey species. In two *Melanocetus johnsoni* were found a single large black lantern-fish each twice to three times the length of the swallower. *A Linopharyne quinaueramosus* contained a large deep-sea eel a hatchet-fish two *Cyclothone* and five 'shrimps' After presenting these and other findings, Bertelsen remarks that though female angler-fishes have diverse kinds of angling devices, there is no evidence of differences in their choice of food. Indeed, as he and Piestch have shown, the luminous complex and dermal decoration of the bait differs from species to species. Acting on conjectures that different kinds of esca might attract different kinds of prey, Pietsch analysed the stomach contents of *Oneirodes* species but found no signs of prey selection. Again it seems clear that there is no choice but the widest possible range of diet in food-poor surroundings.

The exponential decline in the food available for production between succeeding links in a food chain is believed to limit the number of possible links to five or six. If this is so, very productive environments should support more links than do impoverished ones. There are certainly five or six links in the euphotic zone of oceanic regions, but how many are there at bathypelagic levels ? If copepods feeding mainly on faecal material, detritus and microorganisms, correspond to herbivorous epipelagic species, than one might say that first level carnivores, in that they may well be largely dependent on the 'herbivores', are other predatory copepods, amphipods, euphausiids and arrowworms.

Second-order carnivores would be prawns and fishes, but are the fishes, for instance, largely sustained by first-order carnivores ? We have seen that they feed on each other and at all levels of the food chain, Trophic rules seem to be overruled in bathypelagic deserts. The predatory design of bathypelagic fishes is very impressive, but how efficient are they as predators ? When on *RRS*

Discovery I unwittingly staged a striking demonstration of the predatory powers of female angler fish which are described elsewhere.

Ceratioid angler-fishes may be kept alive for sometime after capture, particularly in cool surroundings. I wanted to test the escape responses of ceratioids, for my guess is that the crossed pair of Mauthnerian neurones and fibres, a system essential in the quick escape movements of fishes, might be better developed in the males. 'When I gently held the tail-fin of a male ceratioid in a pair of forceps he wriggled hard as though to escape. Using thumb and forefinger, I did the same to several females but they promptly turned round and tried to bite me. In discussion, Mr. Peter David said that the quick turn-round of female ceratioids reminded him of Mr. Nubar Gulbenkian's taxicab. Mr. Gulbenkian once remarked that his taxi could turn round on a six pence, "whatever that may be". I thus came to call the negative escape response of female deep-sea angler-fishes the "Gulbenkian reflex". Indeed, the females of most species have relatively short luminous lures, a globular deep-bodied form and fins nicely related to the centre of gravity, features that are admirably fitted to a qui-ˡ- rounding on their prey.

After all, a female cannot always expect prey to swim both to wards her flashing lure and straight for her large and gently smiling jaws. Prey may approach from behind, where, incidentally the body is studded with free-ending lateral line organs, able to detect near by disturbances in the water. Studies of the organization of bathypelagic fishes and the 'Gulbenkian reflex' of female ceratioids led me to Sir D' Arcy Thompson's view on deep-sea fishes. In his book *Growth and Form*, Sir D' Arcy remarks. The great depths of the sea differ from habitations of the living, not least in their eternal quietude.

The fishes will dwell therein are quaint and strange; their large heads, prodigious jaws and long tails and tentacles are, as it were, gross exaggerations of the common and conventional forms. We look in vain for any purposeful cause or physiological explanation of these enormities; are left under a vague impression that life has been going on in the security of all but prefect equilibrium, and that the resulting forms, liberated forms, liberated from many ordinary constraints, have grown with unusual freedom.' Contrary to the thought behind Sir D' Arcy's last sentence, the more one looks the more one comes to appreciate how cunningly and closely angler-fishes are designed for life in their deep and difficult environment. It is true that bathy pelagic animals do not face the constraints imposed by vertical migrations. They live in virtually changeless physical surroundings, and they are thus freed from developing physiological and biochemical means to cope with fluctuating conditions. But they are not liberated from one ordinary constraint, the continual need to make a living in impoverished circumstances. Indeed, their life in the most deserted part of the ocean as will now be argued, is ingrained in their whole organization.

8

FREE FLOATING ORGANISMS

The word 'plankton' was coined by Viktor Henson in 1887. Broadly defined it includes all living organisms both plant and animal that drift about passively in water. Many of these organism have the ability to Swim, but because they are small in size generally, progression through the water is relatively slow generally, progression through the water is relatively slow and their distribution is often patchy and determined by the movement of ocean currents and tides. Many planktonic organisms are microscopically small, but there is great variation in size. The pulsating bell of some of the larger jellyfish may be 2 m across with tentacles up to 20 m; these form part of the megaplankton. A more 'biological' classification divides plankton into phytoplankton (drifting plants) and zooplankton (drifting animals). Holoplankton and meroplankton respectively are terms used to describe organisms that spend either their whole lives or only part of their life histories (usually the larval stages) drifting passively in the water. Planktonic larval stages provide a means of dispersal before the adults settle down.

Nets

How is this unevenly distributed plankton caught? The simplest kind of plankton net is conical in shape, with a collecting vessel at the far end. As the net is pulled slowly, so the water passing through is filtered and any plankton caught on the meshes is swept into the collecting vessel. The mesh size will determine if the catch is largely macro or microplankton. Pores soon become clogged in very fine mesh netting and little water will be filtered because the bow wave created will push plankton in the path of the net to one side. There are other problems with this simple kind of net including taking a sample at a particular depth.

The net may be weighted to hold it down, but it may sample not only at the required depth but also on the way down and up. There is no easy way of telling which specimens were collected at different depths when this mixed sample is being sorted in the laboratory. One means of overcoming this problem is by using a Nansen-type closing net. When the net has sampled at the required depth a 'messenger' is sent down the towing warp. This changes the weight of the pull from one bridle to another and the net is closed prior to hauling up to the surface, In order to be certain that the net is closed on the way down and on the way up a Clarke- Bumpus sampler is used. A metal disc sealing the entrance to the net may be rotated by sending a messenger; once the sample has been taken a further messenger causes the entrance to be closed by rotation of the disc. These flap- valves are restricted to nets less than 50 cm diameter for mechanical reasons. Acoustic command signals are now often used instead of brass 'messengers'.

Table 8.1. Size Range of Planktonic Organisms

0.2 - 2.0 m	=	megaplankton (e.g. large jellyfish, fish*)
2-20 cm	=	macroplankton (e.g. some shrimps, fish)
0.2-20 mm	=	mesoplankton (e.g. copepods)
20-200 µm	=	microplankton (e.g. protozoa)
2-20 µm	=	nanoplankton (e.g. autotrophic flagellates)
< 2 µm	=	picoplankton (e.g. bacteria)

The amount of water that flows through the net is difficult to estimate, especially once the meshes of the net become clogged and a bow wave is set up. However, by fitting a flow meter at the entrance to the net and by relating the size of the opening of the net and the reading on the flow meter, estimates of the volume of water sampled may be made by multiplying the mouth area of the net by the distance towed. This is essential if any quantitative estimate is to be made of the amount of plankton in the sea, Non of these refinements is necessary if only a quantitative sample is required, One disadvantage of all these nets is that they may be towed at a few knots only, otherwise the water pressure brusts them. In order to speed up sampling, Sir Alister Hardy devised a continuous plankton sampler which could be used at speed and towed behind commercial ships on their normal routes. It has a

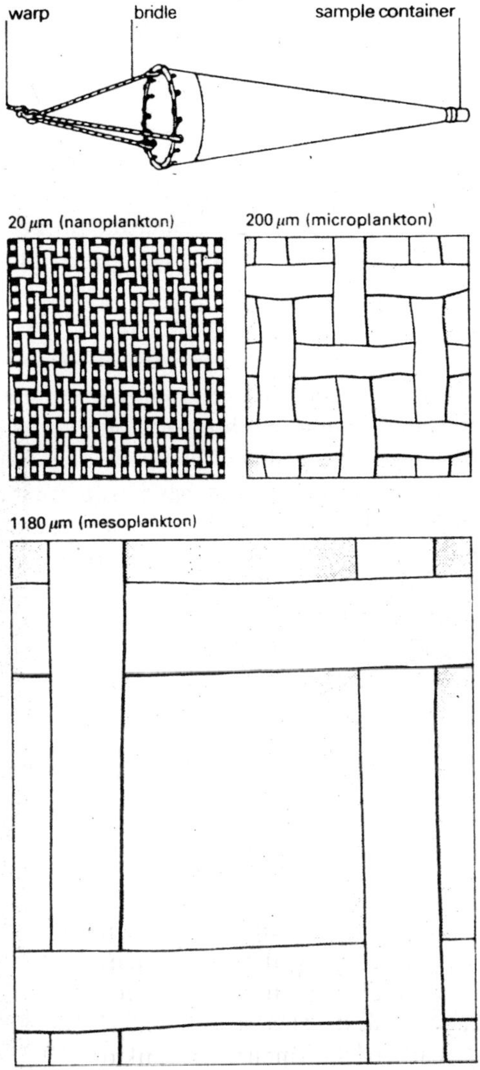

Fig. 8.1. Conical plankton net (top). Types of monofilament polyester bolting cloth showing relative range of mesh sizes.

strong metal outer case and its streamline shape cuts down water pressure.

A criss-cross pattern of cruises built up a good record of plankton, distribution, especially in the Atlantic and in the North Sea. A forerunner of the Hardy sampler was the 'plankton indicator' used by herring drifter skippers to locate waters rich in the copepod Calanus on which the herring fed. High speed sampling of plankton is possible using the Gulf III net made of corrosion resistant cupronickel alloy (hence the abbreviation TTN meaning 'tin tow-net'. The rigid outer casing allows the inner net of metal gauze to be better supported and take the strain imposed at high speed. A nonporous mouth reduction cone increases filtration efficiency by drawing a column of water wider than the reduced mouth, through the sampler

Much emphasis is being placed currently on remote sensing. Infrared photography from satellites, coupled with computer enhancement of images, has enabled the distribution to be plotted of phytoplankton blooms often associated with frontal systems between water masses. Plankton is not confined to the open sea. It occurs in shallow water and even surf. A two-jered plankton sampler towed by hand in the shallows at the water's edge collects plankton washed close inshore. Static plankton nets may be used to collect organisms emerging from the sediment and performing vertical migrations into the water column. Pumping water through filters may be used also to sample plankton in the top room.

Planktonic Organism and Larval Stages

Bacteria

These are certainly to be found, but not usually in very large numbers in the water column because of the paucity of solid objects on which to settle. In a few localities such as in areas of upwelling off the coast of north-west Africa or the hydrothermal vents associated with the spreading of plates farming the earth's surface, high bacterial counts have been recorded in some cases over a billion per litre. The Galapagos rift in the Eastern Pacific is a result of such spreading. A combination of the minerals in solution and the suspended bacteria give a milky blue tinge and shimmer to the hot water ejected from the vents in the rift. Chemosynthetic bacteria in the vent water utilize hydrogen sulphide as source as energy. This contrasts with the more usual dependence of food chains on solar energy.

Phytoplankton

This comprises all those-tiny single-celled organisms containing chlorophyll and other pigments which like land plants, are able to utilize carbon dioxide, nutrient salts and the energy of sunlight to produce carbohydrates, proteins and fats. Most photosynthetic activity occurs in the 15 m of the sea but some occurs even at 100 m in clear oceanic waters. There are two main groups of phytoplankton: diatoms and flagellates.

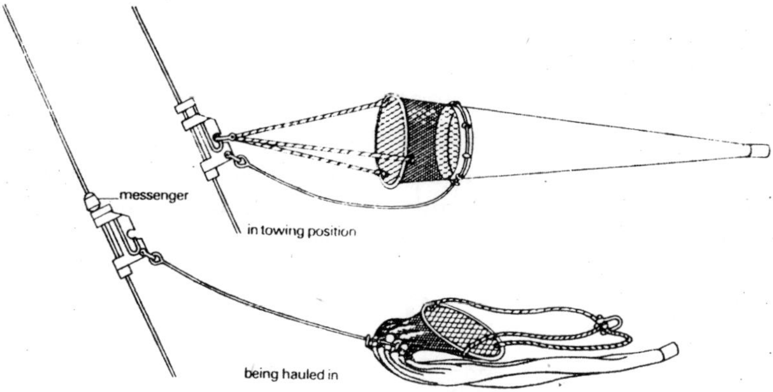

Fig. 8.2. Closing plankton net.

Diatoms

These live mainly in the surface waters and their silica skeleton is often elaborately ornamented which many have the effect of slowing down rates of sinking. Some contain a drop of oil which improves buoyancy. Reproduction is by simple asexual cell division. Each diatom 'shell' is made up of two halves and when the cell divides each half makes another to complete the outer skeleton. After several generations certain of the diatom 'halves' become smaller and smaller but before becoming infinitesimally small, the process is halted by resorting to sexual reproduction. The cell contents of two diatoms emerge from their silica 'shells' after the nucleus has undergone a reduction division halving the number of chromosomes. A nucleus from each partner fuses and produces a nucleus with the full complement of chromosomes, The protoplasm of this newly formed zygote swells and secretes a silica outer skeleton of normal size.

Flagellates

These are more mobile and use their whip-like flagella to provide propulsion through the water. It was thought that all life on land and in the sea depended ultimately on the ability of plants to 'fix' sunlight energy by means of the process of photosynthesis and that these diatoms and flagellates existing in countless millions in the surface waters were the primary producers, providing energy for the entire marine food web. However, chemosynthesis by autotrophic bacteria may play a role in powering some local deep sea communities. Dinoflagellates with red pigments occasionally undergo vast population explosions known as blooms which produce 'red tide'. The dinoflagellates can be removed.

Micro-Zooplankton

In oceanic waters ciliated protozoans of the micro-zoo-plankton consist largely of tintinnids (subphylum Giliophora, order Tintinnoida). Each individual is housed in a case of tectin. These

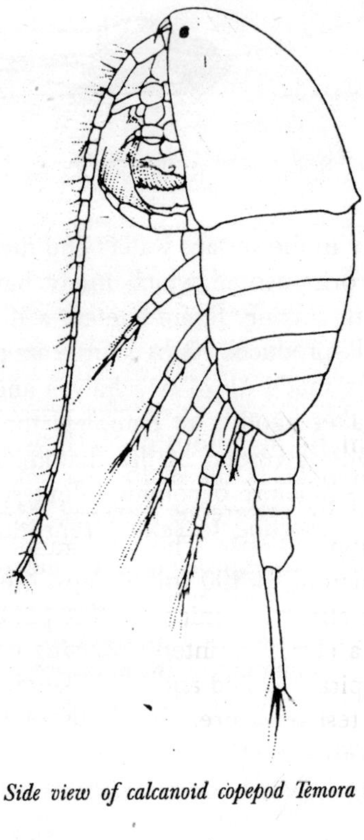

Fig. 8.3. Side view of calcanoid copepod Temora longicornis female.

cases, which are secreted by the animals and vary in form from urn-shaped to tubular, may be plain or more often strengthened by agglutinated material, such as sand grains, diatom valves or coccoliths. Tintinnids range in size from 20 to 640 microns and most species live in warm temperate or tropical regions. At times they become very abundant in coastal waters and the surface layer of the open ocean. Their main food, which consists of organic particles, bacteria and the smallest species of the phytoplankton, is wafted to them in currents created by their apical complex of cilia. 'The remarkable evolutionary versatility of the foraminifera has enabled them to colonize virtually every kind of marine environment with the result that fossils foraminifera are more nearly universally present in the marine sediments of the last 100 million years than any other group of fossils'.

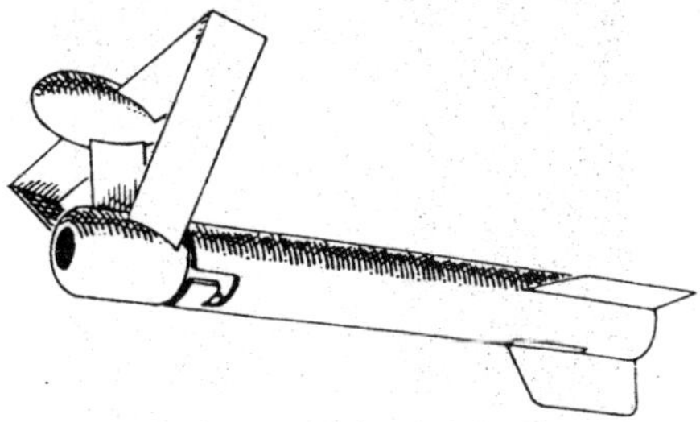

Fig. 8.4. TTN Gulf net for high speed plankton sampling.

Living foraminifera are known from more than 1,200 species, of which thirty or so form part of the plankton of oceanic waters: the rest live at all depths on the ocean floor and will be considered under benthic animals. All species secrete a chitinoid test around due main body of protoplasm. The test is pierced by one or more holes through which the photoplasm flows out to form the filamentous pseudopodia. In the planktonic species and others the test is strengthened by calcareous material. Many of the benthic species have agglutinated tests. Though foraminifera range widely in size from 20 microns to 190 millimetres, the mean size of the common species is about 500 microns. The planktonic species live in tests formed of a cluster of interconnecting chambers, which in many species are spirally coiled and look like miniature snail shells. Beside the main test aperture, the walls of the chambers are punctured by many small holes through which emerge the pseudopodia. When fully extended, these cytoplasmic filaments not only radiate in all directions, but flow into one another to form a network.

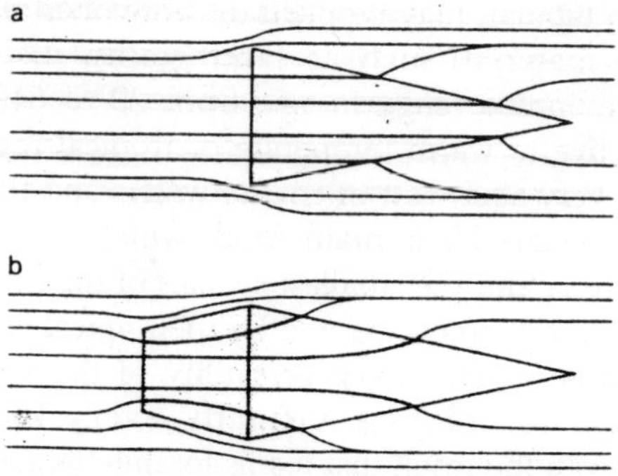

Fig. 8.5. Water flow patterns associated with a simple conical net (a) and a conical net with a non-porous mouth reducing cone (b).

It seems that the main food of the surface-dwelling forms consists of phytoplankton and small ciliates, which are trapped by the pseudopodia and digested on the spot. Planktonic foraminifera are most abundant in the epipelagic zone, where most, if not all, species are though to spend their early life. The epipelagic species generally have spiny tests and the spines not only support an outer, frothy cytoplasm but presumably help to keep the rate of sinking to a low value. The species without a spiny test have a greater depth range and some (e.g. Globorotalia crassiformis, G. scitula and Hatigerinella digitata) are evidently mesopelagic or bathypelagic forms.

Most species live in tropical and subtropical waters, which overlie most of the vast expanses of calcareous sediments that are rich in the shells of planktonic foraminifera (globigerina ooze) and cover about 45 percent of the deep-sea floor. Living Radiolaria, which are known from about 4,400 species, are exclusively marine and with a few exceptions planktonic. The very numerous fossil species go back to Precambrian times. In nearly all living species the endoplasm is housed in a central capsule : the ectoplasm contains, or is contained by, gelatinous layer. As in foraminiferans, long and fine pseudopodia radiate outwards and finally anastomose to form an elaborate food trapping net. Some species are luminescent. All but a few species secrete some kind of skeleton formed of spicules that may be isolated but are much more frequently joined in diverse and beautiful forms, many of concentric and radiating patterns. D'Arcy Thompson was reminded of snowflakes.

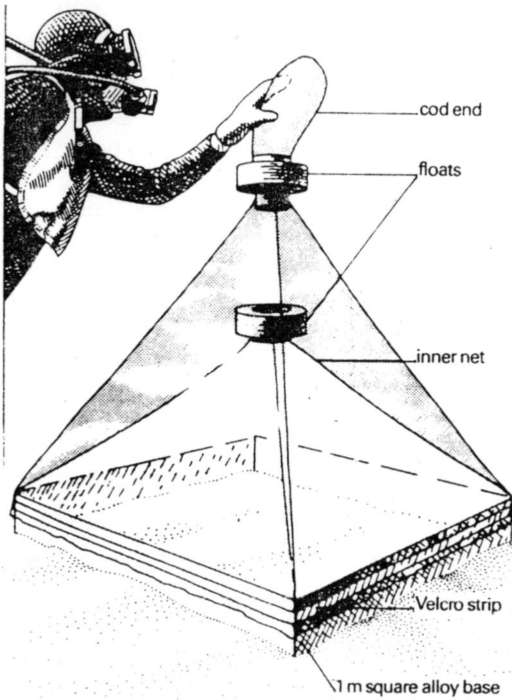

Fig. 8.6. Static plankton net for use in shallow water. Temporary plankton emerge from the sediment, often at night, swim up through the inner funnel and are trapped in the upper portion of the net. Diver recovers net and tows it through water to wash plankton into the narrow cod-end

The little skeletons remind us of snow crystals (themselves almost endless in their diversity), rather than a collection of animals constructed in accordance with functional needs and distributed in accordance with their fitness for particular situations. The unity in diversity of radiolarian skeletons that so intrigued Sir D'Arcy was rediscovered by Buckminster Fuller, who was charmed to find geodesic structuring in their skeletons, and in those of diatoms. Indeed, the skeletons of numerous radiolarians look very like the geodesic domes designed by Buckminster Fuller and others. As he wrote : Employing the fundamental principles of nature's structuring, the average of all the geodesic domes to date enclose space at approximately one percent of the weight of materials required by conventional architectural and engineering structures of equivalent volume designed for equivalent load stresses and clear span environment controls.

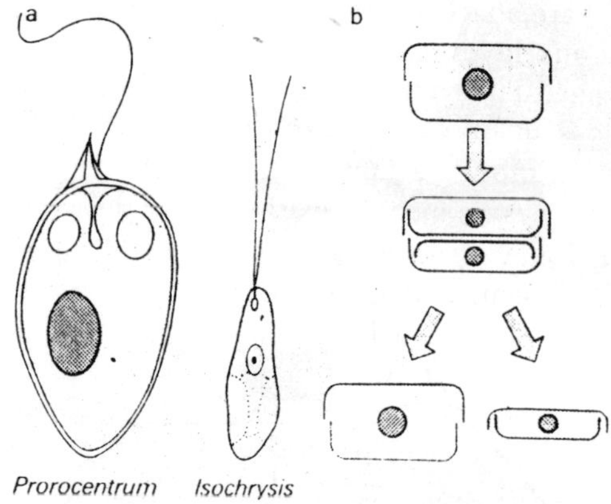

Fig. 8.7. (a) Phytoplanktonic flagellates. (b) Asexual cell division in centric diatoms showing reduction in size of the daughter cell produced by the lower half.

Upheld by the sea, radiolarians face no such stresses, but they presumably need to be supported by skeletal designs that make the best possible use of rare supplies of building material. Curiously enough, the skeletons of the Acantharia (sometimes classified as a separate order), which use rare strontium in the form of sulphate, look more like snow-crystals than geodesic domes But the radiating configuration of spicules is integral to their organization : the spicules support a well-formed gelatinous layer that surrounds the central protoplasmic body. In most acantharians contractile threads

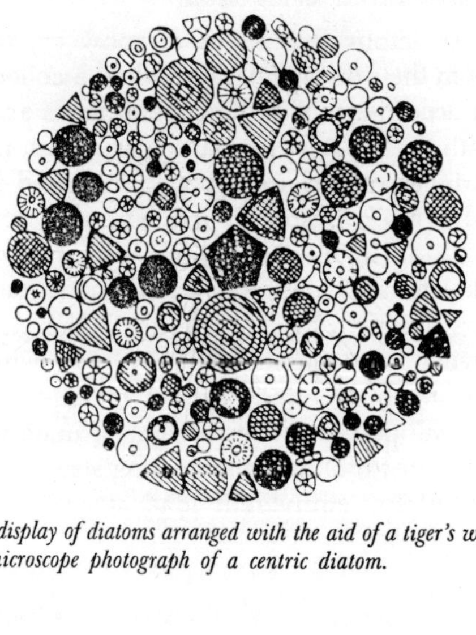

Fig. 8.8. Victorian display of diatoms arranged with the aid of a tiger's whisker and scanning electron microscope photograph of a centric diatom.

(myonemes) are attached between main spicules and the surface of the gelatinous layer, the entire system forming an 'appareil hydrostatique'. The skeleton of the three other main groups of radiolarians is made of silica. The spicules are isolated in certain species but in most are united, often in geodesic array, to form concentric lattices or shells of extraordinary delicacy and diversity. A central capsule is always present.

The ectoplasm, part of which may form a frothy layer that may help in flotation usually contains a well-developed gelatinous layer. The species of two orders, like the acantharians, are predominantly epipelagic. The former are very diverse and include many species with concentric, latticed shells. The deep-sea species belong to a third order, the Phaeodaria, notably to the families Challengeridae and Tuscaroridae. Species of the first family, which have an oval or lenticular shell pitted with minute holes, live at mesopelagic and bathypelagic levels.

In species of *Challengeria* the size increases with their depth range. The tuscarorids, with a spherical, elliptical, fusiform or pyramidal shell, are also small forms with a wide depth range.

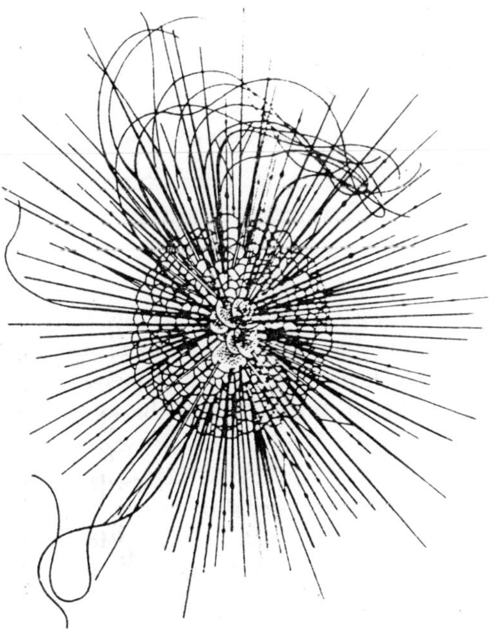

Fig. 8.9. Globigerina in life with spiny spherical chambered shell.

Like the foraminifera, epipelagic radiolarians trap and digest microscopic plants and animals in their pseudopodial complex. They are said to take flagellates, diatoms, tintinnids and even young stages of copepods. In species other than phaeodarians the protoplasm contains the resting stages of dinoflagellates known as zooxanthellae. The latter are photosynthetic and there is presumably some form of nutrients exchange between plant cells and their hosts. Radiolarians are also most diverse in subtropical and tropical regions. At times they are very abundant in productive areas but little is known of their overall success in oceanic waters, During a crossing of the western North Atlantic catches of radiolarians were generally less than those of foraminiferans.

The formation of the siliceous oozes rich in radiolarian skeletons reminds one most of Rachel Carson's evocation of the 'long snowfall' that forms the deep-sea sediments. 'When I think of the floor of the deep sea, the single, overwhelming fact that possesses my imagination is the accumulation of sediments. I see always the steady, unremitting, downward drift of materials from above, flake upon flake, layer upon layer-a drift that has continued for hundreds of millions of years, that will go on as long as there are seas and continents.' It seems odd that radiolarian ooze is almost entirely restricted to the deep-sea floor below the productive waters of the North Pacific Equatorial Divergence. There are also substantial deposit of radiolarian ooze on deep stretches of the ocean floor below the rich waters of the Peruvian Current. Why is so little radiolarian ooze produced in the other oceans?

THE ZOOPLANKTON

Some biological categories are more arbitrary than others, but if the micro-zooplankton is taken to include organisms that will pass through a net with a mesh size of 500 microns, the net, if large enough, would retain all the adult forms of zooplankton, which range in size from 300 microns copepods to the jelly fish *Cyanea capillata*, with an umbrella that the tentacles of a Cyanea may extend for 50 metres or more, the size spectrum of zooplankton covers more orders of magnitude than that of the nekton, with a range from a squid or fish of a few centimetres to a 35 metre blue whale. In taxonomic terms the zooplankton virtually consists of the pelagic forms of Cnidaria, Ctenophora comb jellyfishes, Polychaeta, Mollusca, small crustaceans, Chaetognatha (arrowworms) and tunicates. Young stages in the life history of these animals also form part of the zooplankton, but are considered elsewhere under early life histories. The concept of plankton, as already stated, centres on organisms that drift in the hydrosphere.

In the ocean the speed of water movements ranges widely from quick-moving surface currents, to the slow circulation of the depths, where velocities may be no more than a fraction of a centimetres per second. Velocities of 10 to 50 cm/sec are very common, which are well above sustained swimming speeds of most forms of zooplankton. Though the zooplankton moves with the sea, and indeed, certain species make reliable indicators of water movements, this passive tendency is transcended by strategic activities whereby any species tends to maintain its population and centres of distribution in a moving ocean. In his comprehensive review of propulsion in aquatic animals Lighthill shows clearly that the predominant and most successful form of locomotion is by undulatory movement.

The smallest forms, such as the flagellates and ciliates, propel themselves by undulation of flagella or cilia. Fishes and whales undulate body and fins to make their way. Anniled and nematode worms throw their bodies into waves. Though propelled by jet reaction, 'the second main alternative to undulatory propulsion in water', squids and certain octopods have fins that are fashioned for undulatory propulsion. In moving through water, small forms, whose relative motion is very slow, are faced almost entirely by the viscous resistance of their medium, which varies as the first power of their velocity : larger, faster-moving forms have mainly to contend with the inertial resistance, the resistance of the medium to displacement caused by a moving body, which is proportional to the square of the velocity The relative effect of these two forms of resistance is expressed by the Reynolds' Number R.

$$R = \frac{LVP}{\mu}$$

where L is some dimension (usually length) defining the size of the body, V is the speed relative to the water is the kinematic viscosity, which about 0.1 in water.

The undulatory mode of propulsion is thus effective over a wide range of R, which is one reason, as Lighthill says, why it is the principal mechanism in the motions of aquatic animals. Jet propulsion is appropriate only in animals large enough to yield relatively high values of Reynolds' Number. In this context it is interesting to survey the locomotion of planktonic organisms. The smallest forms of phytoplankton (nanoplankton) are predominantly flagellates, while the rest are also flagellates or non-motile. The main groups of micro-zooplankton are either ciliates or without means of active movements (foraminiferans and radiolarians).

In the zooplankton, however, the small crustaceans and many of the gelatinous forms, which, as we shall see, are dominant groups, have other than undulatory means of locomotion. The copepods, meaning oar-footed, are well named, while the much larger cnidarians and the largest forms of tunicates (Thaliacea) have evolved gelatinous bodies that are shaped to muscular means of jet propulsion. Even so, the cnidarians usually start their free-swimming life as small ciliated larvae (planulae) and some of the tunicates have a larval stage with an undulatory tail. The two dominant groups of undulators in the zooplankton are the smallest tunicates (Larvacea) and the arrow-worms. If biological success is judged by a combination of biomass (the weight of organisms in a given unit of living space), ubiquity and diversity, which in a full sense involves not only the number of species but also the unrivalled in the zooplankton. After the small crustaceans, we shall consider the gelatinous forms of zooplankton (cnidarians, ctenophores and tunicates), and the arrow-worms, which is varying respects meet some of the above criteria. The polychaete worms and molluscs are much more successful as benthic forms.

Small Crustaceans (Copepoda and Ostracoda)

Like their arthropod relatives, the insects, spiders and so forth the Crustacea have segmented bodies and paired, jointed appendages that are covered by a skeleton made largely of chitin and protein. The body is divisible into head, thorax and abdomen. In common with most arthropods, the crustacean appendage is essentially a jointed system of hollow levers that are moved by internal strained muscles. Crustaceans two readily distinguished from other arthropods in developing two pairs of antennae on the head. Close behind the mouth is a pair of mandibles followed by two pairs of maxillae. Copepods have a dorsal carapace that extends over the head and one or two thoracic segments. The first pair of appendages on the thorax (maxillipeds) are succeeded by five pairs of biramous swimming legs in the free-living forms. The abdomen is without appendages. There are about 6,000 species, most of which are marine. Individuals of the pelagic species (about 750) range from about 0.5 to 17 mm in length and live in virtually every part of the ocean Diversity is greatest in the warmer regions. Much more often than not, copepods dominate the zooplankton.

For instance in standard net samples taken during a cruise between New York and Bermuda the volume of epipelagic copepods (measured by displacement) averaged 51 percent of the total volume of zooplankton. Copepods are also likely to be prominent in catches at mesopelagic and bathypelagic levels. Most species are virtually transparent, but when their tissues become cloudy after preservation, they look like small grains of rice. Some of the deep-sea species are tinged in varying degrees with red pigments; a few are black. In some species there are luminescent glands in the skin such that the extrusion of their products causes flashes of light. When copepods are cruising the five pairs of biramous thoracic limbs, which are flattened, paddle steadily backward in sequential strokes. If the animals are disturbed, these limbs beat strongly backward in rapid succession : at the same time the antennae are folded backward along the body. The result is a relatively long dart, and for human eyes at least, such movements in transparent copepods are over in a virtually undetectable flash. The.long leap of Gaussia princeps, a black, deep-sea copepod that teaches a length of just over 10 millimetres, is quite startling.

Copepods of the neuston may even jump out of the sea. Wherever they are, these sudden bursts of speed must often enable copepods to evade their many predators in the zooplankton. Most of the pelagic copepods (calanoids) have long and well-developed first antennae, which may be suddenly flexed to jerk the animals forwards. When copepods are feeding these antennae are extended outward, and no doubt help them to find their food. Species that feed mainly on phytoplankton are generally the most abundant. The classic instance is Calanus finmarchius, which in north temperate and subarctic waters forms much of the food of large whale, bone whales and herring.

Herbivorous copepods filter their food through the long fine bristles of their maxillipeds but if the phytoplankton cells are large enough, they are seized one by one. There are also omnivorous and carnivorous species. Members of all three trophic groups exist together at surface and deep-sea levels of the ocean. In the tropical ocean the numbeı ᴗf copepods in cubic metre of water falls from about 15 in the surface waters to about 2 at lower mesopelagic levels (500 -1,000 m). Between 1,000 and 8,000 metres the number ranges from .42 to 0.05. The considerable success of some mesopelagic species is no doubt related to their daily vertical migration in search of better food supplies above the daytime levels of their populations. Ostracods, which have much the same size range as copepods, are very differently made. The indistinctly segmented body in housed in a laterally compressed, bivalved carapace that may be closed by an adductor muscle. There are two pairs of thoracic appendages but the antennae provide the main propulsive power. Most of the more than 2,000 species are marine and benthic in habit.

Ostracods rival the copepods in ranging adaptability, but the planktonic species are not so abundant. At depths below 1,000 metres the numbers of copepods in a cubic metre are generally much more than ten times those of ostracods. Some of the deep-sea species are luminescent and red or brown in colour. The largest form, *Gigantocypris*, which an almost globular carapace that attains a length of 1 cm, is a bright red *Gigantocypris* normally propels itself with a pair of fan-like antennae protruding through a slot in the carapace at the anterior of the animal.

Balanced strokes propel the animal forward and movement of one or other antenna individually serves to rotate the animals round its vertical axis'. Most planktonic forms are eyeless, but the eyes of Gigantocypris look like '.... two huge forward facing, headlamp-like reflectors, which perhaps assist in its capture of the small fish, chaetognaths, and copepods found in its stomach'. If pelagic ostracods are generally carnivores or scavengers it is natural that they are less numerous than copepods. Compared to the herbivorous or ominivorous copepods their place in the food chain is less close to the primary source of food, the phytoplankton. Like the copepods, some of the ostracods undertake daily vertical migrations. At one station in the eastern North Atlantic, species of *Conchoecia* especially *C. curta*, formed most of the food of the hatchet-fish, Argyropelecus aculeatus (Merrett and Roe, 1974).

Gelatinous Forms

The gelatinous form of zooplankton are cunningly fitted for a marine existence. Though usually jelly-like and fragile, the bodies of cnidarians, ctenophores and tunicates are supported and kept in shape by the sea. In cnidarians the outer and inner tissue of the body wall are separated by a gelatinous layer, which is greatly developed and shaped as a matrix for the parachutes, bells and other parts of planktonic medusae, jellyfishes and siphonophores. Such organization is found also in due comb-jelly fishes. In pelagic tunicates, gelatinous tissue is a matrix for the 'house' of larvaceans, the tunic of salps and doliolids and the tubular setting of pyrosomids. The water content of gelatinous tissue in cnidarians is 95 percent or more.

The matrix contains collagenous fibres that not only support it but give the jelly elasticity as well as a relatively high viscosity. In tunicates the test, whatever its consistency, is made largely of a cellulose called tunicin. More generally, gelatinous tissue forms a hydrostatic skeleton that is very well fitted for propulsion by jet reaction. After each pulsation of the bells of jellyfishes and siphonophores, which forces out a jet of water. The elasticity of the jelly and the return flow of water restore the bell to its resting shape in readiness for the next pulse. Much the same is true of salps and doliolids in motion. When the gelatinous bracts of a siphonophore the detached from the colony they float upwards. They are positively buoyant, but how is this possible in tissue that seems to consist almost entirely of water and sea salts? At Villefranche in 1957 E. J. Denton and T. I. Shaw found that gelatinous animals (like the large dinoflagellate *Noctiluca*) achieve their buoyancy by excluding some of the heavy ions of sea water, notably sulphate, from their body fluids and replacing these ions by lighter ions such as chloride. Denton :

The results of the Villefranche experiments were quite clear. For five kinds of gelatinous marine animals, medusae, ctenophores, tunicates and heteropod and pteropod molluscs, the whole animals were usually either close to neutral buoyancy or positively buoyant. This buoyancy could be quantitatively explained in terms of the salt composition of the animals' body fluids. These were always isomotic with seawater but contained less sulphate and more chloride, and also generally less magnesium and calcium but more sodium than seawater. In all of the animals studied the buoyancy in seawater mainly depended on excluding some sulphate from the body fluids and replacing this anion by chloride.' Presumably gelatinous tissue contains cells that exchange heavy for lighter ions. Whatever the mechanism, the energy needed for such differential ionic transport must be easily saved by the energy not needed, as it were, by buoyant gelatinous animals to keep their level in the ocean. Indeed, the gelatinous nature of these planktonic animals must be seen as integral to their whole organization, which is relatively simple and easy to run.

Most of the cellular elaboration of a cnidarian resides not so much in their means of locomotion but in their means of procuring and digesting food and in reproducing themselves. Peter David's splendid photograph in Deep Oceans of a deep-sea jellyfish with a lantern-fish ensnared in its tentacles is a striking illustration of a simply organized animals availing itself of rich and long-lasting nourishment in a highly organized one. The nervous system and muscles of the prey alone, which was just starting to be digested and is about as long as the bell diameter of the predator, are many times the biomass of the simple nerve net and muscle cells involved in the movement of the jelly fish and its tentacles.

One much not, of course, forget the outlay of materials needed to build the millions of stinging cells in the tentacles (though nematocysts once formed need no maintenance, for they are dead motile units). Even more than their bottom-dwelling relatives, the cnidarians of the plankton are the simplest, deadliest and most effective predators in the ocean. Tunicates are just as cunningly contrived. They are surely the simplest and most effective herbivores in the plankton. As in cnidarians, the feeding, digestive and reproductive systems rather than the nervous and locomotory systems, subsume most of the cellular organization of a tunicate. By means of a house with a system of fine filters (Larvacea) or by means of a mucus filter net secreted by the endostyle (Thaliacea), tunicates are able to trap the smallest and often very abundant members of the phytoplankton. Such plant food is beyond the filtering capacity of most forms of crustacean herbivores, and the larger species are also available to the tunicates. Living organisms range from simple to very complex forms, but how is one to express their degree of organization? Some measure may be got by estimating cellular quantities of deoxyribonucleic and (DNA), which contains the genetic blueprint of all levels of organization in living forms.

The more complex the organism the greater is the amount of DNA. At extreme end of the scale, a human cell contains over two million times as much DNA as that is a simple virus. Sponges have about ten times as much DNA as bacteria. Cnidarians have about six times as much as sponges and in the present context, about twice as much as tunicates, which is remarkable considering that tunicates are related to the choradate ancestors of vertebrate animals. Evidently, DNA content may be a measure of the degree of organization, which is usually but not always related to position in the evolutionary scale. Simple organisms are easier to run than more complex forms of about the same size, and one way of expressing this difference is by measuring their rate of oxygen consumption.

In the jellyfish Aurelia one such measurement is 0.0034 and in a tunicate (Salpa) 0.008, which are among the lowest figures for marine animals. This is not altogether surprising. The jellyfish, for instance essentially consists of simple layers of living tissue dispersed over a massive gelatinous skeleton that is dead apart from its content of amoeboid cells. Moreover, the tissues of a jellyfish or a salp are so thin that they can obtain all their oxygen by simple diffusion: they have no need of respiratory organs.

As Schmidt-Nielsen writes, 'Organisms that are supplied with oxygen by diffusion only, protozoans, flatworms etc. are mostly quite small, less than a mm or so, or have very low metabolic rates, as jellyfish do. Although some jellyfish can be very large, they may contain less than 1% organic matter, the rest is water and salts. They have a very low average rate of oxygen consumption, and the actively metabolizing cells are located along the surfaces where diffusion distances are relatively short.' One can begin to see why planktonic tunicates and cnidarians are sometimes able to flourish in the deep blue deserts of the subtropical ocean. Thus, by a consideration of all the above attributes, the gelatinous forms of zooplankton are well fitted to their (predominant) place in the economy of oceanic life. But any attempt at a full appraisal must include some consideration of their size.

In one respect, as implied earlier, the size range of swimming bells conforms to the nature of the medium in that jet propulsion is effective only when principally resisted by the inertia of water. In the oceanic zooplankton the species with the widest size ranges are gelatinous and jet-propelled, which would still be true if one judges such range only on jet dimensions. Even so, the aspect to be considered here is total size. The greater the size, and the cnidarians also have batteries of noxious deterrents, the less available they are to predators. Predatory neighbours in the plankton can only attack a jellyfish or a salp in piecemeal fashion. To feast on jellyfish the predator must have a mouth as large as that of the ocean sunfish (*Mola*). Salps and medusae are sought by various kinds of large turtles. When the gelatinous forms of zooplankton evolved the means of making and shaping large amounts of their matrix, a wide range of size become possible, yet another biological advantage of their of organization.

Cnidaria

Cnidaria, from the Greek *cnides*, 'nettle', is an apt group name for hydrazoans, jellyfishes, sea-anemones, corals and so forth. There are about ten thousand living species, nearly all marine, and the key to their success and carnivorous way of life resides in the myriads of stinging capsules that form in the outer tissue layer. While large forms, as we have seen, are able to trap, paralyse and feed on sizeable fishes, the overall success of the cnidarians would surely have been much less had they not evolved the right kind of nematocysts for securing the abundant supplied of proteins etc. below the armour of the smaller crustaceans. In Life in the sea by Gunnar Thomson (1971) there is a striking illustration of a small hydroid (*Hydractinia*) swallowing and entirely digesting a copepod, which is about as voluminous as the relaxed hydroid, in the space of 26 hours.

Copepods are strong and active for their size and must be poisoned and paralysed before they can escape. Paralysis requires the injection of toxins below the cuticle of the prey. In a fish, on the other hand, the sensitive epidermis lies exposed over the scales and bones. Nematocysts, the capsules of which range from ellipsoidal to globular in form and in size from 5 to 1.000 microns, are set on the tentacles and certain other parts of cnidarians. Each capsule contains a coiled-up thread with very thin walls. When stimulated, and some nematocysts are triggered by a sensitive hair, the thread is forcibly and very rapidly everted. There are at least three main types of nematocysts : those that coil tightly and cling to the bristles on the surface of small crustaceans; those that fire a lassoing thread, stick or covered with barbed spines; and those that pierce as they explode, so injecting the sting of their contents. The threads of the last type are barbed to the tip and they rotate as they explode. Rotation of the tip and they rotate as they explode. 'Rotation of the tip must confer lateral and longitudinal stability on the advancing threads; and it is striking to see how straight are the tracks of discharges in distilled water, though the discharged tube is a most delicate structure.

Rotation also imparts a cutter-like action to the barbs at the tip, as they pass to the barb series of the everted tube: this is probably unimportant in the lassoing type of capsule, but must determine the cutting and cuticle piercing powers of the injection type of stinging capsule' (Picken, 1957). There is much we do not know of the part played by nematocysts in the life of cnidarians, but we can at least begin to appreciate why they have held their place in the ocean from Cambrian times. We have seen elsewhere that the outer tissue layer (ectoderm) of a cnidarian is separated from the inner layer (endoderm) by the gelatinous layer (mesogloea).

The endoderm lines the body cavity and contains the digestive cells. During development the mysteries of morphogenesis are such that these three body-wall layers are fashioned around the sea-filled body cavity into medusoid or polypoid forms. The adaptations of the gelatinous medusoid form for a planktonic existence should now be evident: polypoid forms, such as hydroids, sea-anemones and corals which have much less mesogloea, are primarily organized for a sedentary life and will be considered later with the other dwellers on the deep-sea floor. But both forms are not only essentially similar in body plan but are also radially symmetrical, particularly in the placing of their tentacles. As Buchsbaum and Milne (1960) say, 'Their elegant symmetry is an effective design for snaring prey from any direction and passing it on to a centrally placed mouth.' The main exceptions to the rule are the numerous siphonophores with tentacles, digestive and reproductive members dispersed along the trailing part of the organism. There are three classes of cnidarians.

In the Anthozoa, which include octocorals, sea-anemones, stony corals and black corals, the life history centres on polypoid phase. The deep-sea forms are considered elsewhere. In the Scyphozoa (jelly-fishes) the medusoid form is predominant: the sessile polypoid stage produces young medusae, or develops directly to the adult condition, or is absent. The groups of Hydrozoa that mainly concern us here are trachyline medusae, the siphonophores and chondrophores. Polypoid members of the important hydroid group live on the ocean floor at all levels. Shallow-sea species may have a medusoid phase (hydromedusae) in the life history. Hydroid

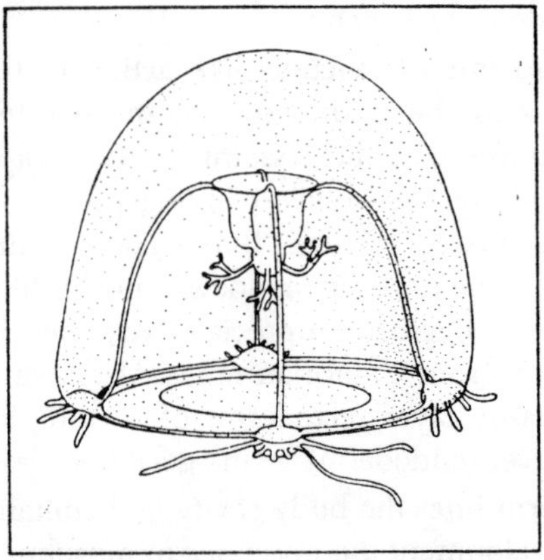

Fig. 8.10. Bougainvillea medusa.

medusae also occur in the deep sea. For instance, during the 39th cruise of the Russian research vessel *Vityaz* in the Kurile-Kamchatka Trench area, such species as *Pandea rubra, Meator rubatra* (Bougainvillidae) and *Aequorea aequorea* (Campanulinidae) were taken. Their relatives in the orders Trachylina and Siphonophora play a more prominent part in deep-sea life.

Hydrazoan medusae, which range from disc-like to bell-like forms, differ from the jellyfishes, inter alia, in developing a velum. The velum is a shelf-like inwards extension of the body wall around the rim of a the medusa so as to restrict its opening to the sea. In the ectoderm of both the velum and the under surface (subumbrella) of the medusa is a circular band of striated muscle fibres, which act together to the shape the swimming movements. Quick contractions of these muscles purse the disc or bell to force water through the yielding and narrowing velar ring in a propellant jet. When the muscles relax, the elastic, viscous jelly and the inflow of water to the subumbrella space restore the medusa to its resting form in readiness for the next pulse.

Swimming movements may be desulatory or repeated rhythmically. The rate of pulsation changes with ambient influences such as the presence of food, the intensity of light and the temperature. Size for size, bell shaped medusae must be faster swimmers than disc-like types, for during pulsations a bell will approach a streamlined form and produce a narrow, relatively

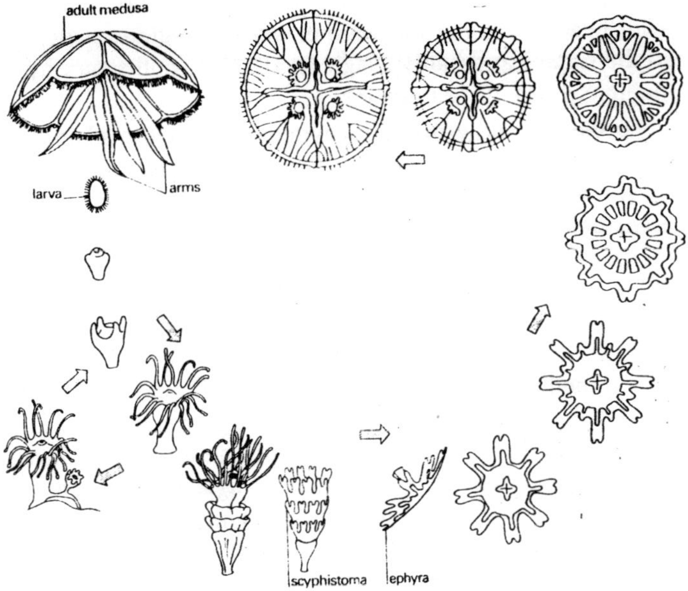

Fig. 8.11. Life history of the cosmopolitan coastal jellyfish Aurelia aurita.

powerful, velar jet. The form and structure of the sea-wasps (Cubomedusae), which are easily the fastest of the jellyfishes, support this contention). The bell-like body, which is disc-shaped in most jellyfishes, has four flattened sides and the margin of the bell is tucked into form a false velum (velarium).

Pulsations may be as much as 150 per minute and speeds of 1 knot or more have been estimated, making the fastest sea-wasps respectable members of the nekton. Thus, a bell-like form producing a concentrated jet and, of course, a well-developed neuromuscular system would seem to make for speed in medusae and jellyfishes. Close comparison of the fishing techniques of bell-like and disc-like species should be interesting. Some disc-like forms (e.g. Cyanea) fish as their tentacles stream outwards during a slow parachute descent. Hydroid leptomedusae tend to be disc-shaped and have balancing organs (statocysts) round margin of the disc.

The other type (Anthomedusae) are bell-shaped with light-sensitive organs (ocelli). Others troll long tentacles as they pulse in a drifting sea. Do the bell types, which presumably rely more on their own movements to carry out their fishing sweeps, catch much of their prey between pulsations, when their tentacles have chance to spread? Most of the trachyline medusae are high-seas forms with headquarters in the warmer oceanic regions at depths between the surface and bathypelagic levels. Bell widths vary from a few millimetres to 10 centimetres.

The species of one group (Trachymedusae) have a plain bell margin and include *Aglantha digitale*, *Halicreas minimum* and *Crossota brünnea*, all virtually cosmopolitan forms. The first, with a transparent bell, sometimes with rosy tints, is most abundant in the epipelagic zone : the populations of the other two, which have bowl-shaped bells, are centred at mesopelagic levels. In Halicreas the radial canals and tentacles are tinged with bright red pigments while the bell of Crossota is darkly coloured, Species of the other group (Narcomedusae have firm glassy bells with margins divided into lobes by the bases of the tentacles. A large round leads to a capacious stomach. Forms that occur at epipelagic to deep-sea levels include representatives of the genera *Aegina*, *Aeginura* and *Solmissus*.

In the Adriatic Sea the daily vertical migrations of Solmissus albaescens have been assessed by over 100 net samples taken between. 1,000 metres and the surface. By day the populations are centred between depths of 400 and 600 metres, and during the afternoon individual start to rise to higher levels. By night if the moon is full the medusae says below a depth of 100 metres but on dark nights swim nearer the surface. This species, which has a well-developed velum with strong circular muscles, was estimated to ascend at a rate of 50 metres per hour, Copepods form the maid food, and the investigator thought that the medusae may have been following their prey. Trachyline medusae which range from a few millimetres to 100 mm in width, are smaller than jellyfishes, but they evidently compensate for their limited prey-snaring radii by relatively high population densities, at least in highly productive waters. In the North Pacific (Kurile-Kamchatka Trench), for instance, maximum numbers of Aglantha digitale were over 1,000 : lower down in the less food-rich mesopelagic zone the comparable figure for Crossota brunneä was 50 (Vinogradov, 1968).

The Siphonophora are pelagic hydrazoans formed of an interconnected complex of medusoid and are polypoid members that serve separate functions and are integrated as a colonial organism. The polypoid members are modified as : digestive zooids (gastrozooids), with a mouth and usually a single tentacle; protective and prey-catching zooids (dactylozooids), slender, mouthless and armed with a tentacle full of nematocysts; and reproductive zooids (gonozooids), producers of the medusoid gonophores that form either eggs or sperm but never develop into complete medusae and are seldom set free. Beside their modification as gonophores, medusoid members form swimming bells (nectocalyces), with velum, four radial canals and ring canals, and bracts–gelatinous, leaf-like appendages that have buoyant properties. It is as though the separate medusoid and polypoid phases of the hydroids have been fused into one polymorphic organism. The 150 or so species of siphonophores are predominantly inhabitants of the oceanic province from epipelagic to deep-sea levels. In two groups (Cystonectae and Physonectae) the colony it topped by a gas-filled pneumatophore or float and they are mainly distinguished by the presence or absence of swimming bells : physonects have one or more swimming bells below the pneumatophore, while cystonects have none.

Members of the third group (Calycophorae) have one or more swimming bells at the head of the colony but are without a pneumatophore. Cystonects include the Portuguese man of-war, *Physalia physalis* a prominent member of the neuston and species of *Rhizophysa. Physalia,* which lives on the warm ocean but may drift into temperate waters, has a sail-topped float below which are tidily attached the tentacular, digestive and reproductive members, In Rhizophysa these members are attached at intervals to a long stem that trails below a sizeable pneumatophore. Nearly all siphonophores are physonects or calcycophorans and most species resemble Rhizophysa in that the tentacular, digestive and reproductive members are hung in a long stem below the head of the colony.

More precisely, the gelatinous bracts cover these members of form separate clusters (cormidia) that are repeat regularly along the stem which is very much alive, especially in its remarkable powers of extension and contraction. Even is species with a float, much of the buoyancy of the colony resides in the bracts. Physonects, such as *Nanomia* and *Halistemma,* draw their line of trailing members through the sea by pulsing a relatively small number (15) of swimming bells. In more complex forms (e.g. *Agalma*) the bank of bells consists of more numerous units. Certain forms (e.g. *Physophora* and *Discolabe*) have the nutritive and reproductive zooids clustered radially about a short stem below the pneumatophore and swimming bells. Nanomia has been well studied. Underwater films from submersible show it capable of lively aquabatics.

Evidently the activities of the various zooids are coordinated by the nervous system. Swimming bells pulse in unison to move the colony along a given line but are equally capable of asymmetrical action to change course suddenly. Contractions of the tentacular members lead to lively activity of the neighbouring digestive members, and in turn digestive products are pumped to other parts of the colonial organisms, surely one of the most fascinating form of animals life. Most species of siphonophores belong to the Calycophorae, some of which are the most widespread and abundant in the order. When mature the cormidia of most calycophorans break loose to become free-living sexual forms, and they are less provalent than the asexual, colonial stages. This is to be expected if the production by gonophores of sexual cells, and their eventual union to form planula larvae, is seasonal. Also the eudoxids are sometimes concentrated at deeper levels than their parent colonies. Species range is size and activity from small darting diphyids to large and sluggish prayids.

The latter forms have two to more large swimming bells that are highly gelatinous and head a long line of trailing members extending for several metres. As meant by their name, diphyids have two swimming bells, which are usually similar and one is just before the other. One species *Chelophyes appendiculata*, may well be the commonest and most widely distributed species of siphonophore. Most siphonophores live in the warmer waters of the open ocean, where numerous species are widely distributed. Species of *Muggiaea* and Nanomia seem to be more neritic than oceanic. The well-known species *Dimophyes arctica* is a cold-water species that lives at epipelagic level in arctic and antarctic areas but is centred in cold, deep waters in warmer intermediate regions. It is one instance of equatorial submergence.

In depth the populations of many species are centred in the epipelagic zone and come extend to mesopelagic levels. Certain species are predominantly mesopelagic and a few are even concentrated at upper bathypelagic depths. Catches off the Canary Island have given intriguing indications of such vertical stratification, especially of how the three most abundant species of *Lensia* seem to have partitioned the water column. Siphonophores, especially the smaller species of diphyids, are among the most abundant of the larger forms of zooplankton. They are so transparent and elusive that it takes some little time before one realizes that a catch is full of the smaller kinds. In their natural surrounding they must be invisible. When they luminesce, a bluish glow lights up the glassy bells of the colony. The purple sail or by-the-wind sailor and its relatives which, like the Portuguese man-of-war, are members of the plueston, were once classified as siphonophores but are now given their own order. The units of the colony hang from a circular or oval, gasfilled float that is strengthened by chitin and in *Velella* bears and upright sail.

There is a large central digestive members surrounded by many reproductive members that produce free medusae and in turn are fringed by a cycle of prey-catching members. Like Physalia, the chondrophores are purple to deep blue in colour and have their sailing headquarters over warmer oceanic regions. Nearly all the deep-sea species of the jellyfish class seem to lack a fixed polypoid stage in their life history. Nearly all belong to the order Coronatae and range from flattened to domelike forms that vary in diameter from a few to 25 centimeters. The ordinal name refers to the groove that crowns the umbrella. Just below this coronal groove the jelly of the bell is fashioned into a circlet of thick pedalia, some or all of which bear a single solid tentacle. The margin of the bell is scalloped into lappets that alternate with the pedalia. As Buchsbaum and Milne say. "The beautiful sculpturing of these masses of jelly remind one of some of the gelatin desserts that have been shaped in grooved, dome-like metal moulds."

The predatory powers of *Periphyllz hyacintha*, which has a dome-like, purple-washed bell and 12 tentacles, has already been stressed. Hardy remarks that this species is usually illustrated '... with its tentacles hanging limply down : in life they are extended in all directions feelings outwards for possible prey'. The saucer-like umbrella of *Atolla* has brown to dark red markings and bears two alternating sets of pedalia with 16-32 tentacles. The disc-like *Nausithoe rubra* is a 'beautiful dark madder-brown'. Representatives of these and other coronate genera are commonest at mesopelagic levels in the subtropical belts but range into colder regions, where they may be taken at epipelagic levels. In the other groups of jellyfishes, *Pelagia noctiluca* with purple-rose lilac colouring, is one of the few species that are truly oceanic.

It is mainly in epipelagic form, though like the deep-sea species, luminescent. Boxjellies and many-mouthed jellyfishes are predominantly inhabitants of warm inshore waters but certain species live in the upper waters of the open ocean. Kramp makes special mention of the 'beautiful brown and violet *Crambionella orisini*, a rhizostome jellyfish that was often taken by the Danish Research Vessel *Galatbea* during her passage from Africa to India. Females of the pelagic octopus, *Argonauta*, the paper nautilus, sometimes attach themselves to this jellyfish. In Kramp's words, between Mombasa and the Seychelles we saw the remarkable sight of number of Argonauta ensconced on top of jellyfish, *Crambionella orsini*, which drifted in large numbers round the ship. To the best of our knowledge this had never been seen before.

Ctenophora (Comb-Jellyfishes)

Comb-jellyfishes are organized in much the same was as cnidarian jellyfishes. The outer and inner tissues are separated and supported by a massive gelatinous matrix that is strengthened by connective tissue fibres and shapes the organism. As in jellyfishes, the meogloea contains amoeboid cells, but there are also muscles fibres, and presumably nerve cells. The two groups are also alike in that the only body space is the endoderm-lined digestive cavity, which has a mouth but no anus. This similarly in degree or organization and the disposition of the body plan in radial symmetry is recognized by placing the two groups in the grade Radiata. The other grade of animals with definite tissues and organ systems, which comprises most of such animals, are bilaterally symmetrical and called Bilateria. More precisely, the ctenophores are biradially symmetrical Radiata.

Most species have mobile tentacles that are densely studded with adhesive cells; cell unique to ctenophores and quite unlike the nematocysts of cnidarians. One ctenophore, *Euchlora rubra*, has nematocysts but no colloblasts. The protruding heads of the adhesive cells are very sticky and cling to prey. At their inner ends they are attached to spirally coiled contractile filaments that yield to the pull of struggling prey but cannot be wrenched loose'. Ctenophores without tentacles just swallow their prey. There are about 80 species, all marine. No ctenophore is built for propulsion by jet reaction most species make their slow way by the sequential beating of ciliary plates that are disposed in radial rows over the animal. Each ciliary plate is made of numerous long cilia that are fused at the attached end and so look like the teeth of a comb, hence Ctenophora, meaning 'comb-cearers'.

The play of light on a moving ctenophore may be the only indication of its presence. As Buchsbaum and Milne say. 'The delicate transparency of comb jellies makes them all but invisible in the water, so that often they reveal themselves only in the rippling iridescence of the rows of beating combs as they diffract the light.' At night, though, ctenophores are brilliantly luminescent, and they flash along the comb rows when distributed by the sea. Lighthill implies that movement by ciliary activity is appropriate only in small organism. At first sight, ctenophores seem to be conspicuous exceptions, for species that move by beating ciliary comb range in size from a few millimetres to about 20 centimetres. First, though, the beating is by sizeable ciliary paddles not by tracts of discrete cilia.

Moreover, the paddles beat in effective metachronal rhythm, more precisely in antiplectic metachronism, which differs from symplectic metachronism in the lag in beat between one ciliary unit and the next is greater, and indeed sufficient to give each paddle greater freedom of movement during its effective stroke. Studies of the ciliated protozoan Paramecium and the ctenophore called the sea gooseberry showed that this kind of metachronal rhythm is more effective than the other is propulsive power. Even so, most ctenophores are slow swimmers. But members of the tentaculate group Lobata, such as *Ocyptosis maculata*, move quickly by strong down beats of their large and wing-like oral lobes. Venus' girdle a ctenophore that is compressed in the tentacular plane to a beautiful band-like form and reaches a length of 1.5 metres, swims by undulations of its body. Apart from a few bottom-dwelling forms, ctenophores are pelagic animals per excellence. Species such as *Beroe cucumis* are cosmopolitan in the widest sense and in cold waters may be tinged with pink or lavender colours. The oceanic species are predominantly epipelagic but there are two deep-water genera.

Tunicate

Much as pelagic cnidarians have evolved, as it were, around their hydrostatic skeleton to become predominant among gelatinous forms of carnivorous zooplankton, so have pelagic tunicates attained comparable success among gelatinous forms of herbivorous zooplankton. There are over 2,000 species of tunicates, most of which are ascidians and benthic in habit from intertidal to deep-sea levels. Since the organization of pelagic cnidarians is easily run, it is understandable that some species live at deep sea levels well below the productive surface zone. Though they may be taken below this zone, and there are a few deep-sea species, pelagic tunicates are primarily forms of neritic waters and the epipelagic zone of the open ocean. Again this might be expected if one considers their dependence on species of phytoplankton. Like their benthic relatives, pelagic tunicates trap such food by means of mucous secretions that come, as already stated, from the endostyle, which is a glandular part of the pharynx.

Larvacea are so called because is certain respects they are like the larval stages of ascidians. Most of the eighty odd species are tiny and transparent with a trunk barely exceeding 1 millimetre, and a tail usually not more than three times the length of the trunk. Special cells on the forward part of the trunk rapidly secrete a transparent and gelatinous filter-chamber (house) that varies in size and complexity from species to species. The house is much larger than its maker, which draws one water through the filters by lashing its tail. In most species of family the water enters through a fine sieve and then circulates inside through a very fine filtering system that funnels to the animal's mouth. Inside the mouth the food particles are trapped in a conical sieve of mucus that is maintained and formed by endostylar and ciliary activity. A twisted cord of mucus and food enters the stomach.

During feeding water pressure builds up in the house and eventually a hinged door by the exit to the house suddenly opens to release a jet of water, and as one jet follows another these extraordinary animals move among the finest pastures of the sea. Species of another family secrete a food trap that is pushed by the animal when it is feeding suspended outside its filter. By such means larvaceans trap the smallest kinds of phytoplankton down to one micron in size. Larvaceans are very common in all parts of the ocean. Species diversity is greatest in subtropical waters. More than is realized at present, these minute tunicates must be an important source of food for young stages of deep-sea animals and for vertically migrating mesopelagic animals, such as lanternfishes. The largest larvacean, *Bathychoradeus*, which reaches a length of more than 77 mm and builds a very voluminous house without filtering windows, is a deep-sea (mesopelagic) form. *Bathychoradeus*, which is believed to feed on detritus and small planktonic organisms, has an enlarged sac-like oesophagus and stomach.

Pyrosomids, salps and doliolids form the other groups. (Thalicea) of a pelagic tunicates. Inside their gelatinous tunic, they resemble ascidians in their general organization, but are modified for pelagic feeding and locomotion in that the atrial cavity is behind the pharynx and discharges water through a posterior opening. Pyrosomids develop into colonies '..consisting of numerous individuals embedded in the wall of a gelatinous tube, which is usually long, more or less cylindrical, closed at one end and open at the other. The zooids have their branchial siphons at the outer surface of the tube, the atrial siphons opining into the common cloacal cavity. Each accordingly passes a water current from the exterior to the interior, where the individual currents fuse into a powerful stream emerging from the open or posterior end of the tube. Locomotion is accordingly a continuous jet propulsion. A diaphragm at the open end narrows the aperture and increases the velocity.

Colonies are often colourless and transparent, but may have a slightly yellowish, blue-green colour. They may be soft and gelatinous or rigid and cartilaginous in texture. The outer surface is never quite smooth'. Though predominatly warm-water forms, pyrosomids are found in all oceanic areas apart from the Arctic Sea. Colonies range in length from a few centimeters to over 4 metres. As their name suggests, pyrosomids are luminescent: indeed, they may well produce the brightest light of all luminous marine organisms. The light comes from photogenic bodies on each side of the pharynx and has been said to range in colour through red, orange, greenish, sky blue, greenish-blue and azure. But one set of careful measurements showed that most of the light energy is blue-green with a secondary greenish peak. Luminescence has also been observed in the larvacean genus Oikopleura and in salps and doliolids. Though they are rather simply organized, doliolids have a complex life cycle. Berrill (1950) remarks that their ...method of attainment of a sexually mature form is probably the most intricate in the whole of the animal kingdom'. This sequence of events is less complicated in salps. Essentially both groups have an sexual and a sexual phase in the life history.

In salps solitary individuals with no gonads (oozooids) have a budding stolon producing a chain of individuals that form the aggregate phase and produce ova and sperm. The fertilized eggs produce oozooids. Doliolids have solitary sexual individuals (gonozooids) and the fertilized eggs produce tailed larvae that grow into asexual oozooids. Each oozooid has a ventral budding stolon and a dorsal spur at the posterior end. Buds formed by the stolon migrate up and become attached to the dorsal spur where they grow into multiple lateral rows of feeding individuals (trophozooids) and a median row of phorozooids that eventually become detached and grow into gonozooids. In the groups and sexual forms have an individual structure, but in essence the means of propulsion and food procurement are similar.

Propulsion is centred in the muscle bands that entirely or partly encircle the body wall just below its substantial gelatinous tunic. As these muscles relax and contrast water is taken into the brachial chamber and expelled in jets from the atrial chamber. Bursts of locomotor activity are interspersed with longer and irregular pauses.

Much as in larvaceans, phytoplankton organisms and particles that enter the pharynx are trapped in a mucous net that is secreted by the endostyle and drawn across the pharynx by ciliary activity. In the aggregate form of the salp *Pegea confoederata*, for instance, the mucous net is conical and continuously renewed and is evidently able to retain particle as small as I micron. The remains of flagellates, coccolithophorids, diatoms, tintinnids, foraminiferans and radiolarians were found in faecal material. Salps are larger than doliolids. Solitary salps vary in length from a few to 200 millimeters.

Chain of aggregate forms may reach lengths of 2 metres or more. Single doliolids span from a few to 30 mm and the chains of zooids on the nurse oozooids reach 40 centimetres. Like other pelagic tunicates, salps and doliolids are mainly cosmopolitan in the subtropical and tropical belts. A number of species extend into temperate and polar waters. In equatorial regions much of the planktonic biomass in the summer months consists of salps, doliods pyrosomids. Sometimes the sea is thick with tunicates and at right the light displays of pyrosomids is a sight not to be forgotten. Apart from turtles, fishes and numerous invertebrates, notably heteropods, their main predators are various kinds of seabirds.

Pelagic Snails (Pteropods and Heteropods)

In two groups of pelagic gastropods the muscular foot, which is one of the key features that contributes to the outstanding success of gastropods in benthic habitats, has been modified into wing-like appendages. Hence they are known as pteropods. More precisely, the wings are outgrowths of the sides of the foot and form the propulsive organs of the shelled pteropods (Thecosomata) and the naked pteropods, which both belong to the opisthobranch group of gastropod molluscs but are not closely related. In one family of shelled pteropods (Limacinidae) the wings and head emerge from a transparent shell measuring a few millimetres in length or width. Morton (1958) writes, 'The foot has a broad sole and bears an operculum; and from its sides are produced long parapodia, or "wings", muscular at the bases and thin towards the tips. *Limacina* rows itself upwards in a broadly spiral course, using these as oars, and drops as a dead weight by holding them together motionless above the body.' Species of the family Cavoliniidae have a bilaterally symmetrical shell, which may be long and conical (*Creseis*), dorsoventrally flattened (*Clio*) or inflated and flask-like (*Cavolinia*), *Cavolinia tridentata*, which is about 22 mm long, bears a pair of greenish, leaf-like appendages, each strengthened by a central vein that emerges from a rear angle of the shell.

Presumably these appendages enable the animal to plane through the water, and not lose too much height, between bursts of swimming activity. Members of another family are large, gelatinous and close to perfect buoyancy. They have no true shell but in a boat-like pseudoconch made of chitin. 'In *Gymbulia* a pair of broad based wings are attached along the sides. They are flapped like a butterfly's wings with up and down rowing movement; and they show also an undulation from the front backwards, which provides a forward thrust.

In *Gleba* and Corolla the pseudoconcha is a wide saucer and the wings have grown together in a heart-shaped fringe which is moved gracefully in swimming like a skirt or the bell of medusa' Morton, 1958. The shelled pteropods feed on microorganisms, particularly on the minute plants of the nanoplankton. Species of Limacinidae and Cavoliniidae set up ciliary current on the wings and mantle, so drawing in organisms and food particles that are trapped by film of muscus. Cymbuliids secrete a large mucus web. Gleba cordata, which has a wing span of about 50 mm, spins a 2 metre-span web by means of mucus glands at the tips of the wings. 'Gently manipulated by the cilia in the animals' elongated mouth, the free-floating membrane slowly sinks, collecting a variety of tiny planketers on its sticky-surface. Eventually, Gleba draws most of the web down its proboscis and consumes it.' Like other members of the zooplankton that are predominantly herbivorous, the shelled pteropods are both common and abundant.

In the Beaufort Sea region of the arctic, the cosmopolitan species *Spiratella belicina* appears to be the cause of a sonic scattering layer at a depth of 50 metres. Nearly all pteropods are epipelagic and most species live in the warmer oceanic belts, where a number are circumtropical. Shells of pteropods may be found in the sediments, but pteropod ooze occurs in significant amounts only in the Atlantic Ocean, mainly in the central South Atlantic. Cymbuliids, with their large digestible cast-nests of mucus, are evidently very efficient herbivores and probably much more abundant than is appreciated at present. The naked pteropods are much less abundant than the shelled species, which is not surprising in view of their carnivorous habits The body is more or less spindle-shaped and may attain a length of more than 40 mm.

Gymnosomes move quickly on powerful wings. 'The animals move upwards or forwards by sculling with them, twisting them at the "wrist" to allow a power stroke at both the upward and downward beat'. The prey, much of which consists of shelled pteropods, is taken by a formidable buccal armature consisting of various combinations of prehensile hooks, sucker-studded tentacles and adhesive oral papillae. One species, *Clione limacina*, which, like its prey *Spiratella helicina*, is sometimes an important item in the diet of baleen whales in arctic waters, is also similar in its 'bipolar' distribution. But gymnosomes are most diverse in warmer oceanic regions. Heteropoda, known by some 25 species, belong to the large prosobranch group of snails. Like many other planktonic animals they are virtually transparent. In one family comprising the smallest species, the coiled transparent shell is so compressed and keeled that the snail stays upright as it swims, which as in all heteropods, is effected by undulations down a fin-like part of the foot. The shell in much reduced in the Carinariidae and entirely absent in the Pterotracheidae. Members of both families have an elongated gelatinous body that is close to neutral buoyancy. When these heteropods swim quickly, which is aided by sepentine movements of the body in pterotracheids, the fin side is uppermost.

All heteropods have large tubular eyes moved by special muscles, which may well help them find their prey. Morton (1958) writes, "The fast swimming Heteropoda, with movable telescopic eyes, are active predators, catching medusae, small fish, copepods, and pteropods and themselves falling prey to larger heteropods. The large *Pterotrachea* and *Carinaria* carry the buccal bulb at the end of their flexible "trunk" periodically thrusting out the sharp-toothed radula to seize prey, or even to attack the human finger-tip. Heteropods swallow their prey entire, which also includes hyperiid amphipods, salps and arrow-worms. The headquarters of these lively pelagic snails is in the epipelagic zone of the subtropical and tropical belts, where numerous species are more or less cosmopolitan. But their place in the ocean is limited by comparison with that of the arrow-worms, which are also predators and have much the same range of lengths.

Chaetognatha (Arrow Worms)

Three groups of worm-like invertebrates are predominant in marine ecosystems. In the zooplankton of both the neritic and oceanic provinces, the Chaetognatha or arrow-worms are much more ubiquitous and abundant than the pelagic species of polychaete worms, which, together with the nematode worms, are outstanding among the worm-like invertebrates in benthic environments. Arrow-worms maintain their place in the ocean, which goes back to Cambrian times, with relatively few (about 50) species. Most are below 40 mm in length and the entire range is from a few to over 100 millimetres. A rounded head, armed on each side with a set of grasping spines, passes evenly into a transparent, torpedo-shaped body that extends in the horizontal plane into one or two pairs of laterals fins and a terminal tail fin. The fins are immobile and supported by a double set of delicate fin rays.

Apart from a few benthic species of the genus *Spadella*, all arrow-worms are planktonic and most species are placed in the genus Sagitta. In most arrow-worms the body is more or less stiff and turgid and is held straight between periods of activity. Movement comes from the quick contractions of substantial bands of muscles that run down the body. Arrow-worms seem to spend much of their time slowly gliding on their fins.... eventually however, they begin to sink, whereupon they resort to swimming to maintain their level. Swimming consists of short swift forward darts, each covering a distance of around 5 cm; this is followed by gliding on the momentum of the darts and a return of floating. The forward dart is so rapid as to be difficult of analysis but is believed to result from the alternate contractions of the dorsal and ventral longitudinal muscle bands of trunk and tail. This would seem to produce up-and-down body waves and tail flirts, but the movements is so rapid that it appears tot he naked eye like a trembling' (Hyman, 1959).

Arrow-worms may dart to elude their predators or take their prey, both of which are detected and located by groups of bristle-bearing sensory cells that remind one of the neuromast organs in the lateral line system of fishes Marshall, 1972. The prey is seized in the spine-like jaws and swallowed whole. Copepods are the most frequent prey, but other chaetognaths, euphausiids, medusae, pelagic tunicates and young fish are also taken. Perhaps chaetognaths feeds on most kinds of zooplankton that they are able to swallow. Unlike most other groups of planktonic animals, chetognaths have no larval phase. They are hermaphrodites and the ovaries mature after the coelomic cavities of the tail are full of ripe sperm. In Sagitta self-fertilization is said to be common.

The eggs hatch into young looking rather like the adults. Wherever they live, and in oceanic waters arrow-worms are world-wide in occurrence at epipelagic and deep-sea levels, fecundity and the cycles of reproduction are such as to maintain relatively abundant population. Between New York and Bermuda the displacement volumes of epipelagic arrow-worms were usually second to those of the copepods and averaged 13 percent to the total volumes of zooplankton. At levels from the surface to 1,000 meters in the western North Pacific the biomass of arrow-worms was about 10 percent of the total biomass of zooplankton and micronekton irrespective of the depth.

In absolute terms the number of individuals–and the number of species– per unit volume of water are greater in the upper 200 meters than below this layer. Species diversity is greatest in warmer oceanic waters. Thus, in the subtropical and tropical waters of the Indo West Pacific faunal province, which extends from the Red Sea and east coast of Africa across the Indian Ocean to the Hawaiian Islands and Polynesia in the western Pacific, about 85 percent of the known species have been recorded, some of which are circumtropical.

Species such as *Sagitta bipunctata, S. enflata, S. bexaptera* and *Pterosagitta araco* are cosmopolitan at epipelagic levels in both temperate and warm oceanic waters. Species with populations centred at mesopelagic levels include *Eukrobnia spp. Sagitta lyra, S. decipiens* and *S. macrocephala. Bathyspadella edentata,* which may be bathypelagic, is without jaw spines. Certain mesopelagic species may be tinged in varying degrees with orange or red pigments. Except for the bluish pigmentation of neustonic forms, other chaetognaths are virtually transparent. Diurnal changes of level associated with vertical migrations are well marked in certain species in neritic waters, where populations tend to be concentrated near the surface during the night. Oceanic species may be partial migrators or non-migrators.

In the Southern Ocean, where *Sagitta gazellae* is one of three endemic species, there was no clear evidence that *gazellae* is a diurnal migrator. But there are seasonal changes of level associated with the life cycle. David's analysis suggests that maturing animals gradually sink and eventually release their eggs at about a depth of 1,000 metres. The eggs rise and probably hatch at about 250 metres, the young arrowing into juvenile animals that mass in the upper 100-metre layer. These growing populations migrate to the 100-200 metre layer in the winter months, but rise and resume life in the surface layer in the spring.

Polychaete and Nemertean Worms

Like the arrow-worms, planktonic species of polychaete worms are predatory and known from relatively few (about 60) species. Most belong to four families (Alciopidae, Tomopteridae, Typhloscolecidae and Phyllodocidae) of which the first two are entirely comprised of pelagic forms and best known to marine biologists. Fauvel (1959 has caught their attraction. '*Les Tomopteris et les Alciopiens pelagiques sont transparents commedu cristal et serainent a peu pres invisibles n'etaient less taches vivement colorees des parapodes des premiers et les enormes yeux rouges des seconds.*' ('The pelagic tomopterids and alciopids are as transparent as crystal and would be almost invisible if it were not for, firstly, the vividly coloured sports on their parapodia and, secondly, their huge red eyes.') Some of the brightly coloured spots on the parapodia of tomopterids are traceable to rosettes of yellow cells that luminesce through large surrounding hyaline cells.

The large projecting parapodia end in paddle-like expansions, and it is by sequential strokes down the parapodial series that (tomopterids row swiftly through the sea. Alciopids have shorter (uniramous) parapodia but have lively motions as they seek their

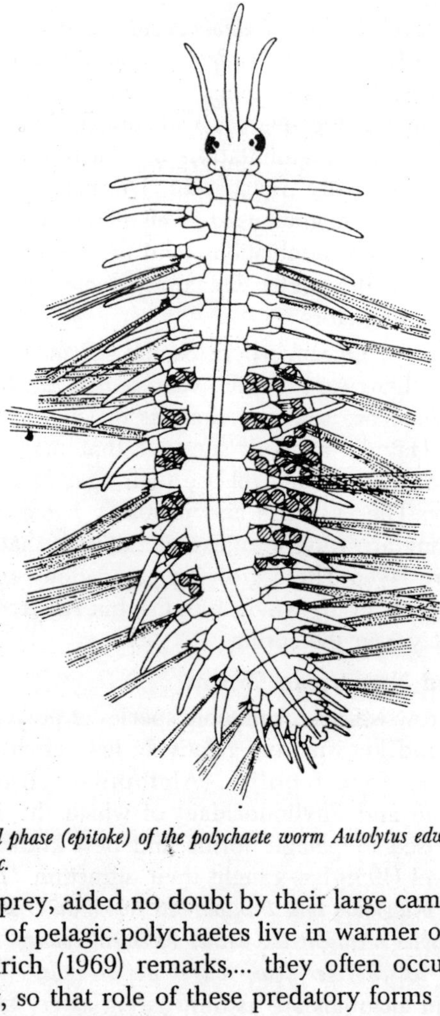

Fig. 8.12. Sexual phase (epitoke) of the polychaete worm Autolytus edwardsi. Female with egg sac.

zooplankton prey, aided no doubt by their large camera -type eyes. Most species of pelagic polychaetes live in warmer oceanic regions and as Friedrich (1969) remarks,... they often occur sporadically and solitarily, so that role of these predatory forms in the biology of the pelagial is of secondary importance in comparison with that of the benthonic polychaetes which play a very important part in the biology of the benthos.' One species, *Tomopteris septentrionalis*, lives at epipelagic levels in arctic and antarctic regions but at deeper levels in the warmer oceanic belts.

Certain deep-sea species of Tomopteris, as Sir Alister Hardy (1956) illustrates so well, are almost entirely red. Some of the pelagic nemertean worms, all of which seem to live at deep-sea levels, also develop red pigmentation. The body, which ranges in length from 6 to 20 cm and is usually widened, sometimes with undulating edges, has a gelatinous mesenchyme. Species of *Pelagonemertes* have a transparent body and a branched, orange or red-coloured gut : others such as *Bathynemertes hardyi*, are scarlet or plum-coloured, *Nectonemertes* has been observed to swim rapidly with strong undulations. Pelagic nemerteans presumably use their long protrusible proboscis to take their prey in the zooplankton.

NEKTON (AND MICRONEKTON)

Nektonic animals have greater swimming powers than those that comprise the zooplankton, and much as one may usefully distinguish a micro-zooplankton from the larger forms of zooplankton so may the smaller kinds of nekton be considered as micronekton. The members of each group have a size and activity range that may be sensed by the dimensions, mesh size, and towing speed of the nets that have been designed to catch them most effectively. A small conical net with a mesh size of 200 microns and towed at a speed of about half a metre per second is used to

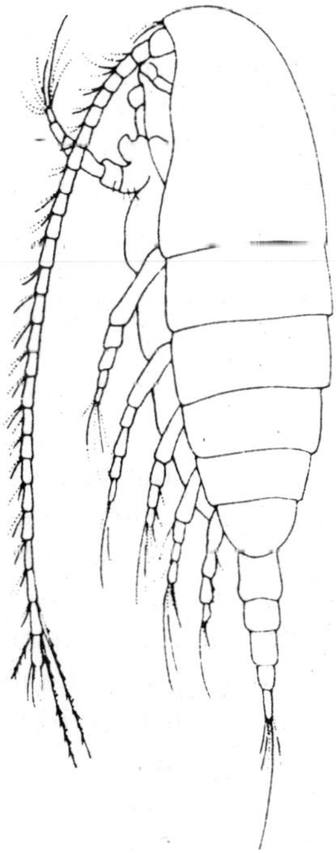

Fig. 8.13. Copepod Calanus finmarchicus.

sample the micro-zoo-plankton. A larger conical net with a mouth area of 1 square metre (70 cm diameter) and mesh size from 300 to 500 microns, which is usually towed at something like 1-3 metres per second, samples the larger members of the zooplankton. The net designed by the Institute of Marine Resources. La Jolla, California, to sample the micronekton of eastern Pacific waters has a mouth area 2.3 square metres and oblong meshes of 5.5 x 2.5 mm. Towed at about 5 metres per second, this net takes the larger crustaceans, small cephalopods and fishes of a size range from about 1 to 10 centimetres (Strickland, 1972).

Nets such as the Isaacs-Kidd midwater trawl and the modified Tucker net used by the Institute of Oceanographic Sciences with a mouth area of several square metres, and smallest (codend) meshes of 77 meshes-inch, are towed at speed of 11/2 knots (0.77) metres per second) or more and catch micronekton and smaller fish and squid members of the nekton (> 10 cm in length.) Still larger mid water nets increase the size range and catches of nekton, but there are still limitations. As Clarke (1971) observes, "Large commercial midwater rawls are regularly used for biological sampling, and here the mouth is held open by otter boards floats on the headline and large weights on the footrope. Examples are the British Columbia midwater trawl, the Engels midwater trawl and the four-door Cobb midwater trawl. These trawls are difficult to operate from multi-purpose oceanographic vessels. However, they are valuable in sampling animals up to about half a metre in length.

The catch is not easy to use in comparative studies because the mouth area and shape change with speed, and graduation of the mesh size along the net makes interpretation difficult. Even with nets of this enormity with mouth areas of 2,800 square metres and 750 square metres (less when fishing because they close up due to drag) the larger animals, known to be present by observation are not caught'. Such observation may be made by examining the food eaten by sperm whales. Elsewhere, Clarke has calculated that a sperm whale of eleven tons eats no less than 130 tons of squid in a years. Many of these squid will be over a metre in length and some, such as *Architeuthis*, span well over 10 metres. Apart from their size and activity, micronektonic animals differ from those of the zooplankton in form and organization.

The shrimp-like myside and euphausiids and the deep-sea prawns, like the squid and fishes, have more or less streamlined spindle-shaped bodies (Marshall, 1971). In keeping with their size and complex organ system, al these animals have elaborate gills. In the twilight world of mesopelagic levels eyes are very well developed and many species have complicated constellations of light organs. Size, form and activity are correlated with massive muscular systems. The joint exoskeleton of crustaceans is associated with a complex system of muscles, which are very highly developed in micronektonic forms. In a mysid, euphausiid or deep-sea prawn the weight of muscle must be well over 50 percent of the total weight.

During cruising movements the most active part of this muscle is that moving the swimming legs (pleopod), but during escape movements the whole of the powerful abdominal musculature propels the animal backwards by a series of rapid tail-flips. Muscle and skeleton make up much of the dry weight of such forms. For instance, the dry weight as a percentage of the wet weight reaches 15 percent or more in amphipods, mysids and euphausiids. In siphonophores, medusae, arrow-worms and tunicates the corresponding figure ranged form 0.3 to 10.0 The mantle muscle of squids, which expresses propellant jets from the mantle cavity; is about 30 percent of the body weight in active species. In active fishes the segmented lateral muscles used in swimming are 40 percent or more of their total weight. Lastly, a high percentage of micronektonic forms are carnivores, whereas many zooplankton species, forming much of the total biomass, are herbivores. In these words, compared to the zooplankton most organisms of the micronekton may be regarded as predatory-behaviour-machines that are quite expensive to maintain. The greatest contrast between the micronekton and nekton is in the epipelagic zone.

Here is the swim of baleen whales, dolphins, turtles whale-shark, basking shark, isurid in gray sharks thresher-sharks marline, spearfish sailfish, swordfish and various squid (ommastrephid and onychoteuthid). This is Isaac's vivid appreciation, 'The pelagic region contains some of the largest and most superbly designed creatures ever to inhabit this earth : the exquisitely constructed pelagic tunas; the multi-coloured dolphin fishes, capturers of flying fishes; the conversational porpoises; the shallow- and deep feeding swordfishes and toothed whales, and the greatest carnivores of all, the bateen whales and some plankton-eating sharks, whose prey are entire schools of krill or small fishes. Seals and sea lions feed far into the pelagic realm. In concert with these great predators large carnivorous sharks await injured prey. Marine birds, some adapted to almost continuous pelagic life, consume surprising quantities of ocean food, diving, plunging, skimming and gulping in pursuit.

Creatures of this region have developed such faculties as advanced sonar, unexplained senses of orientation and homing, and extreme olfactory sensitivity.' Compared to all these large and active members of the nekton, the crustaceans and small fishes of the micronekton are indeed shrimps and so much whitebait. One need only contrast this state of affairs with conditions immediately below in the mesopelagic zone which is the most productive part of midwater space, to see how much life in scaled down in the deep sea. If the size range of micronektonic animals is taken to be about 1-10 cm, then virtually all of the larger crustaceans and many of the cephalopods and fishes are included. Some of these animals, such as certain species of euphasuiids, prawns, squid and lantern-fish. Swim up to the epipelagic zone each day, where they form part of the micronekton during the night. The nekton of the mesopelagic zone consists of a few large species of fishes and squid. Though more is known now of the larger forms of pelagic deep-sea animals, Hjort's description of this fauna as Lilliputian still seems apt.

Euphausiacea

If their near ubiquity, trophic range and numerical abundance are considered together, the euphausiid shrimps emerge as the copepods; of the micronekton. Considering just the oceanic species, euphausiids live from epipelagic to bathypelagic levels throughout the four oceans. Species fall into three main trophic types, as in copepods herbivores (the conditions are favourable), omnivores and carnivores. In one measure of abundance, the mean displacement volume of euphausiids was about 5 percent of the total catch figure in a series of net hauls in the epipelagic zone between Vew York and Bermuda (Grice and Hart, 1962). As already stated the corresponding figure for copepods was about 50 percent, and one must always remember that copepods are much easier to catch than euphausiids. At levels between the surface and 1,000 metres from stations in the western North Pacific, the biomass of euphausiids was 8 to 15 percent of the total catches of zooplankton and micronekton. During the antarctic summer, vast swarms of euphausiids, nourish baleen whales, crab-eater seals, penguins and fishes. The 85 species of euphausiids (order Euphausiacea) are closely related to the large order of decapod crustaceans which includes shrimps, prawns crayfish, lobsters and crabs.

Indeed, certain authorities class them with the Decapods. All species are shrimp-like in appearance and hence well-developed antennules and antennae that bear olfactory and tactile receptors. Just behind the mouth the paired mandibles, maxillules and maxillae collect, macerate and manipulate the food just before it is swallowed. The outer branches of the 6-8 pairs of thoracic limbs, to which the gills are attached, produce currents of water from which food particles are screened by the setae on the inner branches (endopodites) of these limbs. Euphausiids cruise by the sequentia paddling motions of the two pairs of swimming limbs on each of the first five segments of the abdomen. The sixth and last segment

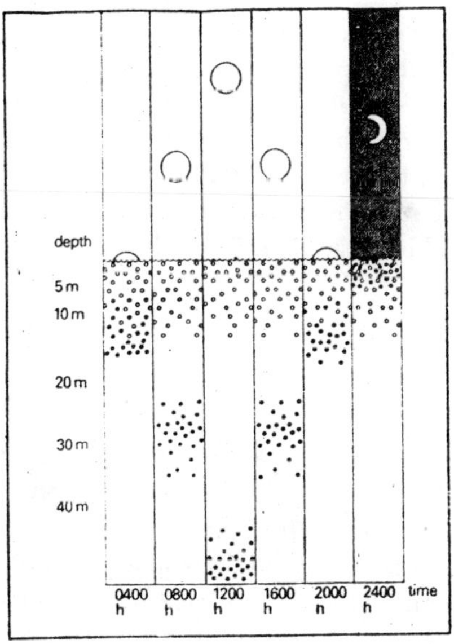

Fig. 8.14. Vertical daily migration pattern of Calanus (open circle–phytoplankton, black dots–Calanus).

has no such appendages but articulated at the end of this segment is a slender tapering telson, on either side of which is a flattened biramous limb (uropod).

Telson and uropods form a tail fan, which is brought powerfully into play by a series of quick flexures of the muscular abdomen, when the animal shoots backwards, perhaps to evade a predator. Species range in length (lip of rostrum to end of telson) from about 10 to 150 mm. The largest of the epipelagic) species is Euphausia superba the whalers' krill of oceanic areas south of the Antarctic Convergence. Small species of Euphausia and various smallish species of other genera (e.g. *Stylocheiron* and *Nematoscelis*) are typical of the warmer epipelagic waters. Other species of these three genera and representative of such genera as physanopoda and Nematobrachion live at mesopelagic levels.

Large species of Thysanopoda (adults of T. cornuta and T, seinicaudata may exceed 100 mm in length) and *Bentheuphausia amblyops*, which are widely distributed, are typical of bathypelagic levels. These species of Thysanopoda may be largely red in colour : other species may be touched and spotted with reddish pigments but are otherwise virtually transparent. Unlike other euphausiids, Bentheuphausia amblyops has regressed eyes and no light organs. Most species have ten light organs: one in each eye stalk, one pair at each of the bases of the second and seventh thoracic limbs and a single organ lying between each of the first four pairs of abdominal swimming legs. Each photophore has a lens and group of light-producing cells. The Antarctic krill Euphausia superba, which feeds

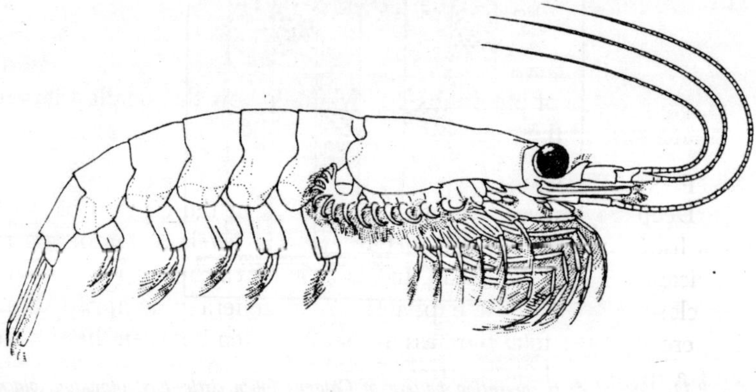

Fig. 8.15. Antarctic krill Euphausia superba.

on the abundant crops of phytoplankton during the summer, is a seasonal herbivore. At other times it is an omnivore like most of the euphausiids.

Carnivorous species, which prey on copepods and other suitable forms of zooplankton, belong to the genera with bilobed eyes: stylocheiron, Nematoscelis and Nematobrachion. These forms also have a very long pair of anterior thoracic limbs ending either in a few large spines or small chelae, presumably for grasping captured prey The functional design of the feeding and alimentary devices of euphausiids and the part these animals play in oceanic food chains and economy are considered later. Each day certain mesopelagic species migrate between their daytime levels and the surface water where they fine much of their food while other species, both epipelagic and mesopelagic, stay at one level. Recent survey show that migrant species of the same genus overlap in their daytime vertical ranges, whereas there is vertical segregation between nonmigrant species,

Euphausiids produce-small eggs and after fertilization each eventually hatches into an ovoid nauplius larva a millimetre or less in length. The nauplius has three pairs of swimming legs which later form the two pairs of antennae and the mandibles. The nauplii develop after moulting their exoskeleton into second stage nauplii. A series of growth stages and moults takes the nauplius through a metanauplius stage calyptopis stages (with a hooded carapace and more definite abdomen) and furcilia stages, which become more and more like the adult. The last furcilia stage moults, forms and changes into an adolescent. Deep sea species apparently prawns at depth and the sequence of events in some mesopelagic species, at least, is that the early larval life is passed in the productive surface water. If all the larval stages of bathypelagic species remain within the depth range of the adults one wonders how the smallest larvae manage to feed themselves.

Deep-sea Prawns

Deep-sea prawns are often prominent in the catches of large nets fishes from epipelagic to bathypelagic levels in temperate to tropical reaches of the ocean. In biomass the catches of such prawns are close to those of the euphausiids and generally comprise about 10 percent of the total biomass in samples taken between the surface and a depth of 1,000 metres. Deep-sea prawns, euphausiids and fishes make up most of the micronekton and their contributions respectively range from 15-25, 35-50 and 20-45 percent of the total biomass. As a rule euphausiids are smaller than prawns, so the former easily outnumber the latter. Moreover, the part played by prawns in the pelagic life of subpolar and polar waters is insignificant compared to that assumed by the euphausiids. Shrimps and prawns form the Natantia, or swimming decapod crustaceans.

Lobsters, cray fish and crabs, which are predominantly benthic. The characteristic features of natant decapods are the body is laterally compressed; the base of each antenna is articulated to a movable plate-like scale, the thoracic legs are slender, and there is a full complement of well-developed swimming legs (pleopods) on the abdomen. In nearly all species the carapace just forwards as a prominent rostrum, which is often saw-toothed and compressed, During quick movements, when prey may be seized, the compression of the body and rostrum will tend to keep a prawn from veering off a line that is guided by the antennal scales, which act as rudders.

Compensatory movements of the scales may also help to control rolling and pitching. Deep-sea prawns thus seem to be well fitted for an active, pelagic life. They also have a much lighter, less well calcified exoskeleton than the bottom dwelling, crawling decapods. The latter never have a laterally compressed body, antennal scales are vestigial or absent, and the pleopods which are not used in swimming, are often reduced. Swimming decapods fall into two main groups, Penaeidea and Caridea, which are readily distinguished.

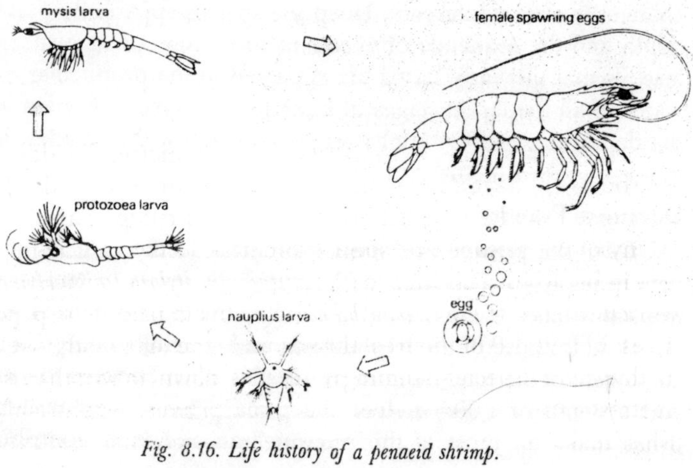

Fig. 8.16. Life history of a penaeid shrimp.

Carideans have expanded sideplates (pleura) on the second abdominal segment that overlap those of the adjacent segments, whereas there is no such overlap is in penaeids. Furthermore, most penaeids have a small claw at the end of the third thoracic leg, which is not developed in carideans. Deep-sea penaeids such as *Gennadas*, *Funchalia* and sergestids have extremely long whip-like antennae and the lash is thickly and regularly set with setae, many of which are presumably tactile or olfactory in functions. These prawns are rather slender in form but they have powerful pleopods and in some the rearmost pairs of thoracic legs have become strong jointed oars. One has the impression of sudden powers to dart on prey detected by the searching sweeps of the long sensory antennae. Deep-sea carideans also have strong swimming legs. Species of the family Oplophoridae are stoutly made: those of *Notostomus* have an 'inflated' carapace. There are over 200 species of pelagic prawns, most of which range from 10 to 100 mm in length. The largest reach lengths from 130 to 180 mm and often seem to live near the deep sea floor (Omori, 1974). Nearly all species live in the subtropical and tropical belts and some extend into temperate waters.

As already indicated, deep-sea prawns are poorly represented in the coldest oceanic regions, where, for instance, in antarctic waters five species have been recorded. There are parallel differences in the latitudinal diversity of euphausiids, but the six antarctic species, especially Euphausia superba, are key forms in the economy of life. Euphausiids are also more prominent than the prawns, both in individuals and species, in the temperate to tropical reaches of the epipelagic zone. Of the mesopelagic prawns, those that live around 500-700 metres or above by day are transparent, or translucent with scattered red chromatophores, or half red (Sergestes spp., Funchalia and Parapandalus).

All species that occur below 500-700 metres by day (lower mesopelagic) have numerous small chromatophores and a heavily pigmented exoskeleton that ranges from deep red to scariet in colour. Many of the mesopelagic species undertake daily vertical migration but in bathypelagic forms (e.g. *Acanthephya stylorostralis*, *Hymenodora glacialis* and *Physetocaris microphthalma*) the main part of the population lives below 1,000 metres throughout the entire day. Certain bathypelagic species seemingly extend down to depths of 5,000 metres or more. Some deep-sea prawns have light organs. The most remarkable are modified liver tubules (Organs of Pesta) which may have reflectors and lenses to enhance and concentrate the down-welling light. These organs are developed in both penaeids and carideans but in the former group, for instance, they occur only in the semitransparent upper mesopelagic species. The lower mesopelagic species of *Sergia* are a uniform red colour and there are no Organs of Pesta.

Dermal photophores may also be present, and those that have lenses also live at upper mesopelagic levels. The all-red species develop dermal photophores without lenses and these live, as already stated, at lower mesopelagic levels, Bathypelagic species, such as Sergia japonicus, are without photophores. There are thus correlations between type of pigmentation photophore development and depth ranges. Indeed, as Foxton has suggested, and is considered elsewhere these correlations remind one of parallel developments in certain cephalopods and fishes.

Deep-sea penaeids produce numerous small eggs that are shed into the water. After fertilization they hatch into minute nauplius larvae with the usual three pairs of swimming legs that later become the antennules, antennae and mandibles. The nauplii change after moults into protozoea stages, which have functional mouthparts. Then comes a zoea stage (called acanthosoma in some sergestids) in which some or all of the thoracic limbs functions in swimming, but the abdominal segments acquire their full complement of setose swimming legs. In some species of Sergestes and Sergia the carapace

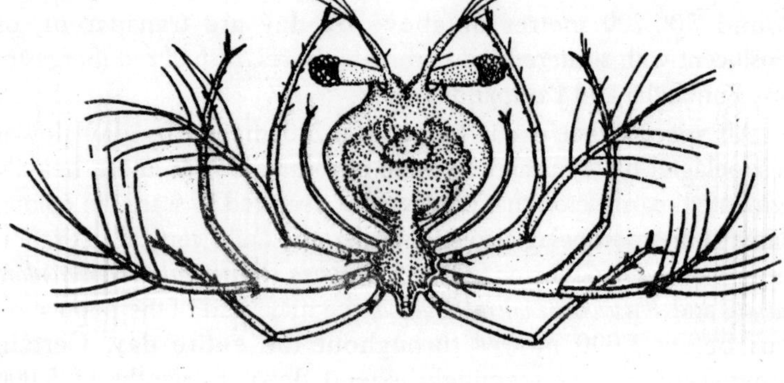

Fig. 8.17. Phyllosoma larva.

of the protozoea and zoea larvae bristle with long spines that may have a protective function as well as helping to slow down the sinking rate of these larvae. Whereas the eggs of sergestids are a fraction of a millimetre in diameter, deep-sea carideans produce relatively few, large eggs which range from about 1 to 4 mm (along the longest axis). The eggs are carried by the female on long setae of the abdominal limbs where they stay until they hatch into relatively advanced stages.

After a few post-larval stages the adult form is attained. The early larval stages are passed within the egg, and as Omori says, the advanced young that are produced have well developed swimming and feeding powers that give them considerable safety from predators and independence of supplies of minute food. The penaeids rely on safety in numbers. A comparative survey of such reproductive and larval strategies in the deep-sea will be found elsewhere.

Peracarid Crustaceans

In euphausiids, shrimps prawns, lobsters, crayfish and crabs, which are classes as eucarids, the carapace is large and covers the thorax. The eggs, if carried, are attached to the pleopods. Peracarids, as implied by their name, have a brood pouch in which the female bears the rather large eggs until they hatch. The carapace, if present,

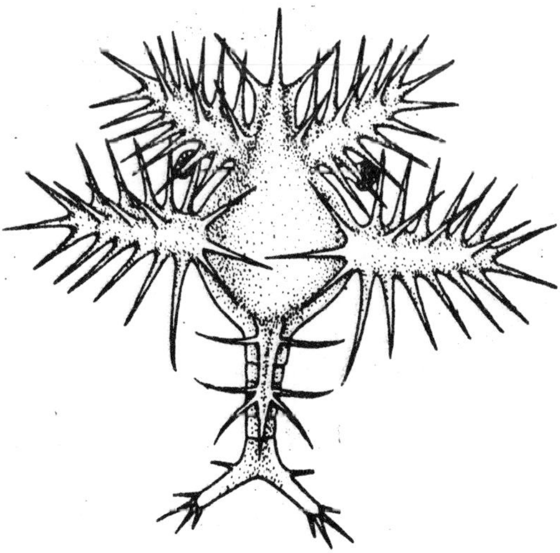

Fig. 8.18. Sergestes, zoea larva, diagrammatic.

is not fully developed and leaves at least four of the thoracic segments free. Pelagic peracarids belong mainly to the opossum shrimp group and the Amphipoda, Cumaceans, tanaids and isopods, which are entirely or largely benthic. Deep-sea myside have been taken in nets fishing below 6,000 metres. But they are commonest at mesopelagic levels, where live species of such genera as *Gnathophausia*, *Eucopia* levels, where live species of Gnathophausia, which are red and look like prawns, are also prawn-like in size.

Most myside are from 20 to 30 mm in length, but a female *Gnathophausia ingens* measured 351 mm (from tip of rostrum to tip of telson) and other large specimens of this species range from 155 to 210 mm. Species of Eucopia, which have long, frail-looking thoracic legs, are also red. There is a fine painting of a scarlet *E. unguiculata* by Sir Alister Hardy Many mysids are filter feeders, but the deep-sea forms are likely to be omnivores. Amphipods have sessile eyes, no carapace, and in nearly all species the body is flattened from side to side and presents an arched profile. Representatives of the subgroup *Hyperiidae* live at epipelagic and mesopelagic levels over most of the ocean. *Hyperia*, *Themisto* and *Parathemisto* species are most common are epipelagic levels, and individuals of one live commensally it seems, with large jelly fishes. The transparent frail-looking Diogenes shrimps (*Phronima*) with enormous eyes on a hammer head, have more drastic designs on gelatinous organisms.

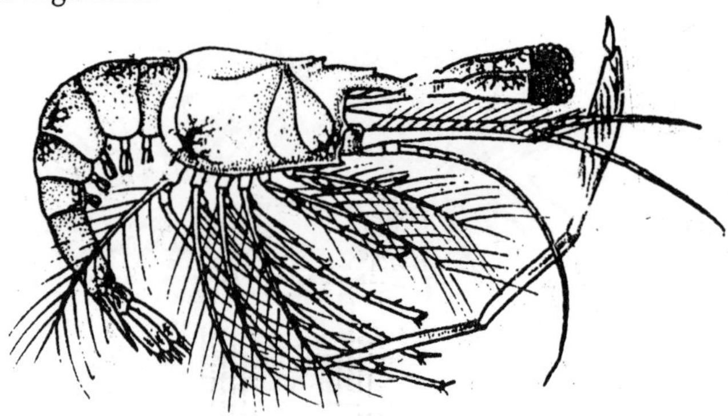

Fig. 8.19. Pandalid, schizopod larva.

Adult Phronima, which may be common at mesopelagic levels and range from about be common at mesopelagic levels and range from about 20 to 30 mm in length, will take over the swimming bell of a siphonophore or the test of a salp or pyrosomid and then proceed to eat away the inside of such structured until a gelatinous barrel is left. The Phronima lives in the barrel and moves it along by limb movements that draw water through its shelter, which also serves as a brood chamber and a nursery for the young. Cystisoma, another fragile but larger (15 cm) amphipod with a near crystalline transparency, is seen from time to time in nets fished at upper, mesopelagic levels. Of one fine photograph, the Angels remark, 'The only colour in this specimen is the interference colour of the eye and some red pigment from the animals' last meal." But certain mesopelagic hyperiids are all red, and again we are indebted to Sir Alister Hardy for fine paintings. Representatives of the largest group of amphipods (Gammaridea) also form part of the mesopelagic fauna. Even so, the peracarid crustaceans are less prominent in the economy of pelagic life than their eucarid relatives, the euphauriids and prawns.

Cephalopoda

When a research vessel in on night station, squids and fishes may be attracted to the illumination cast into the sea by the deck lights. The squid may also be drawn to the fishes, though it is difficult to tell predator from prey. One sees dark spindle-shaped bodies and glinting eyes moving just below the surface. Even in brighter surroundings a darting squid may be momentarily mistaken for a fish. Through such illusions an observer is made a ware of the extraordinary similarities between cephalopods and fishes. Beside

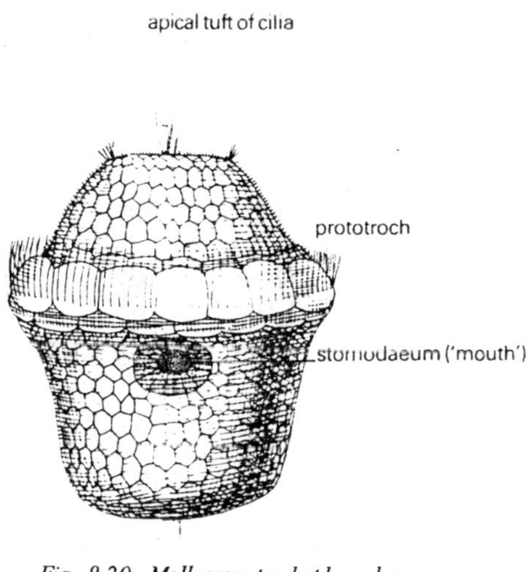

Fig. 8.20. Molluscan trochophore larva.

the streamlined shape of active species and their prominent camera-like eyes, the two groups resemble each other in numerous other respects of their gross and find structure. These resemblances form just one striking illustration of the brute fact that during their evolutionary history distantly related kinds of plants and animals have animals have converged in functional design in response to the pressures of natural selection in a given medium, or to subdivisions of a medium ranging from broad adaptive zones to narrow ecological niches, Though this is not the place to consider convergence, it must be said that this aspect of evolution is prominently displayed in the deep sea. Many of the 700-800 species of modern cephalopods live in the deep sea.

Pelagic species are mostly squids together with a number of octopods, a few sepioids (cuttle-fish) group and the vampire squid. Most of the octopods and seploids are bottom dwellers. Epipelagic species, notably the ommastrephid squid, have firm torpedo-shaped bodies and strong swimming powers, which reside in their thick, muscular mantles. One species and a hooked squid, which have large triangular fins, have power enough to dart out of the water and fly through the air. After take-off, flying-squid are sustained largely, it seems, by continued expulsion of jets from the funnel, and they may cover distances of 50 metres or more.

When a small shark approached two *Ommastrephes*, both were observed to turn a dark red and shoot out of the sea to a height of 2 metres. Though primarily epipelagic Ommastrephid squid have been photographed by baited cameras at mesopelagic levels. As in fishes epipelagic species are considerably larger than most of their relatives of lower levels. The flying-squid (*O. bartrami*) reaches a length of about 1 metre, while *Dosidicus gigas*, an ommastrephid of Chilean waters, grows to well over 2 metres. There are smaller and more delicate squid in the epipelagic zone, especially members of the family Cranchiidae. Species of *Cranchia* and *Liocranchia* which may be quite abundant, have soft, globular bodies, large eyes, short arms and rather long tentacles.

Apart from certain internal organs, these squid are virtually transparent. The liver is a particularly opaque organ, but in Liocranchia it is silvered over, presumably to hide it from predators. Cranchia scabra attains a body length of about 50 mm, which is less than half the length of an adult *Locranchia reinhardti*. Concerning the cephalopods of deeper levels, closing mid-water nets fished accurately from RRS Discovery, the research vessel of the Institute of Oceanographic Sciences, UK, have enabled Malcolm Clark and his colleague C. C. Lu to determine the vertical ranges of the commonest North Atlantic species throughout their growth stages and where possible to trace the daily changes in level of the species that undertake vertical migrations. They also found a latitudinal gradient in species diversity. From 60° N to about 20°N the number of species taken roughly doubled for every ten degrees of latitude. Thus, like other groups of marine organisms, cephalopods are most diverse in the warmer parts of the ocean (Clark and Lu, 1974 and 1975; Lu and Clark, 1975). Much of what follows is taken from their work.

Mesopelagic squids of the upper 700 metres '.. are in general small squids rarely exceeding 30 cm in overall length excluding 30 cm in overall length excluding the tentacles and averaging only about 15 cm with large numbers adult are $^{4\text{-}5}$ cm'. In the North Atlantic the commonest small squid of mesopelagic levels include *Abraliopsis affinis*: *Histioteushis* spp.; *Pyroteuthis margaritifera* and *Mastigoteuthis schmidti*. In Abraliopsis and *Histioteuthis* most of the numerous light organs are on the lower side of the animals. The few jewel-like light organs of Pyroteuthis are set around the eyes, at the tips of its tentacles and in the body. Indeed, an outstanding feature of mesopelagic squid '... is the possession of numerous small to large light organs or photophores distributed chiefly upon the vertical surface of the mantle, funnel, head arms and the ventral periphery of the eyeball'. Many mesopelagic fishes also have numerous light organs along their underparts, and, as in squid, some of these organs along their underparts, and, as in squid, some of these organs are of elaborate design.

Again, mesopelagic squid and fishes have large and very complex eyes, but no fish has eyes such as develop in histioteuthid squid, which have the left eye much larger than the other. Two small seploid squid, *Spirula* and *Heteroteuthis*, are also common in parts of the Atlantic at mesopelagic depths. Spirula floats head downwards, buoyed by an internal, gas-filled spirally chambered shell, and shining upward at the end of the body just above the shell is a light organ. Spirula is the only deep-sea cephalopod with a gasfilled buoyancy system. In diverse squid neutral buoyancy is achieved by buoyancy chambers filled with relatively light ammoniacal liquids. Such a system in found also in certain deep-sea prawns. Fishes manage well, if need be, with a gas-filled swimbladder.

The other sepioid, Heteroteuthis, looks rather like its bottom-dwelling relatives. The ink it discharges on being disturbed is mixed with a luminous secretion, which reminds one of the clouds of luminous material emitted by certain deep-sea prawns and fishes. As already implied, convergent evolution is rife in the deep sea, especially in visual, luminescent and buoyancy systems. In the catches from their North Atlantic stations, the two commonest midwater octopods studied by Clarke and Lu were *Vitreledonella richardi* and *Japetella diaphana* which both grow to a mantle length of about 60 mm.

Both species are virtually transparent and have the gelatinous consistency and something of the appearance of a jellyfish. The fist is mesopelagic at all stages of growth, whereas half-grown and adult *Japetella* were taken at upper bathypelagic levels. A third gelatinous mid water octopod. *Amphitretuspelagicus*, has upturned tubular eyes that are reminiscent of similar organs in certain mesopelagic fishes. The finned octopods, such as *Cirrothauma* and *Girroteuthis* once thought to be bathypelagic, have been photographed near the deep-sea floor. They are also gelatinous and the large webs between their arms make them look even more like jelly fishes.

The paddle-like fins project from a relatively small body. Cirrothauma is blind and certain species of *Cirroteuthis* have regressed eyes. No deep-sea octopod, or indeed any kind of octopod, or indeed any kind of octopod, is know to be luminescent. The vampire squid, *Vampyroteuthis infernalis*, which is more or less cosmopolitan between lower mesopelagic and bathypelagic levels, is neither a squid nor an octopus. With its eight broad-webbed arms, paddle-like fins and soft gelatinous body it looks rather a like finned octopod, but there are certain outstanding differences. Actually the vampire has ten arms, two being modified into long, mobile sensory filaments that are withdrawable into special pockets.

Thus, it is also quite distinct from the squids in which two of the ten arms develop into special, prey-catching tentacles. Vampyroteuthis is placed in a separate order called the Vampyromorpha. It is also unlike any octopod in its well-developed set of light organs. Besides the many small organs that stud the skin, there are two clusters of luminous nodules on the back of the 'neck' and two large lights, each with an 'eye-lid' just behind the fins. Males reach a length of about 125 mm. females about 200 mm. Some kinds of mesopelagic squid migrate daily between their daytime levels and overlying waters, where they swim by night, and presumably find much of their food. Among the species that are most common the North Atlantic, Clarke and Lu found reasonable evidence that *Abralliopsis affinis*, *Bathyteuthis abyssicola* and Pyroteuthis margaritifera appear at epipelagic levels during the night.

Other common squid, such as *Heliococranchia pfefferi* and Mastigoteuthis schmidti, are non-migrators, as are the octopods Japetella and Vitreledonella. There are also ortogenetic migrations during the life history. The catches from Discovery stations suggest that most deep-sea cephalopods start their young life in the epipelagic zone and assume deeper levels as they grow to maturity. There is uncertainty as to where the eggs are laid. In 1895, when the *Princesse Alice I*, the research yacht of Prince Albert 1 of Monaco, was working near the Azores the local whalers harpooned a sperm whale, which happened to die under the bottom of the yacht. In its death flurry is spewed up parts of great squids new to science.

On returning to Monaco the Prince had his yacht fitted as a whale-catcher and began a series of investigations on the sperm whale and its food, showing that it feeds mainly on squids, some with tentacles as thick as a man's arm and studded with great suckers bearing catlike claws. But for these and subsequent investigations, and for the occasional stranded specimen, our knowledge of squids, particularly of their size range, would be very restricted. Sperm whales feed on medium-sized to the largest squid, which must be more abundant and larger than appears from the cathes of midwater nets.

Fishes

Small fishes, euphausiid shrimps and deep-sea prawns make up most of the micronekton. Why are small squids not a major constituent, or are sampling devices at fault? Perhaps squids are the hardest to catch, through the elusive powers of a fish would seem to be much the same as those of a squid of the same size and shape. But the smallest pelagic squid are larger than the smallest fishes. Most of the prolific species of mid-water fishes, which nearly all belong to the stomiatoid and lantern-fish groups, range from 25 to 70 mm in length when mature. If the arms are included in the length, hardly a small sized species of squids falls into the above size span, which suggests that the smallest squids are more elusive than the smallest fishes. But are the latter more productive than the former, bearing in mind that animals feeding nearest the primary producers in a food chain are the most productive. Are small fishes better able than their squid counterparts to take advantage of the wealth of small (mainly copepod) food in the ocean?

Small stomiatoids and lentern-fishes certainly feed on eggs, larval stages, copepods, ostracods, arrow-worms, euphausids and so forth, but less is known of squid food. Prey caught by a squid or octopus is conveyed by the arms to the beak-like jaw and there chewed, if necessary, into pieces small enough to enter the oesophagus, which is restricted by its passage through the brain and cephalic cartilages. Diverse predatory fishes are able to swallow prey up to half of their own size. Small species of midwater squid, including migratory species, also feed on copepods and euphausids and certain mesopelagic and bathypelagic cephalopods have suckers small enough to grip small planktonic animals. Further observation and investigation is needed. In the meantime one suspects that the fishes are able to take a wider range of small prey organisms, and thus feed closer than the cephalopods to the photosynthetic base of the food pyramid.

So instance, it is improbable that any genus of deep-sea cephalopod has species to match the abundance and ubiquity of those in the stomiatoid genus Cyclothone. The most diverse mesopelagic fishes, both in numbers of species and individuals, are stomiatoids and lantern-fishes. There are over 300 species in the second group and about 250 in the first. Both groups have a few representatives at bathypelagic levels near 1,000 metres and below, where, indeed most fishes belong to the dark-coloured species of Cyclothone. But in diversity alone the bulk of bathypelagic fauna consists of ceratioid angler-fishes, which are known from about 100 species. Stomiatoid fishes, which seem most closely related to the smelts, typically develop four rows of prominent and complex light organs along the underparts of the head and trunk and two rows at the same level along the tail.

Small silvery kinds, such as *Vinciguerria*, *Valenciennellus* and *Bonopartia* Gonosstomatidae) and some of the hatchet-fishes (Sternoptychiade) are common at upper mesopelagic levels, but they are usually outnumbered by small transparent fishes, again of the genus Cyclothone. The smaller species of stomiatoids have relatively long jaws bearing numerous pointed teeth, part of a prey-taking system to deal with the range of food organisms already listed. Most of the larger species belong to the families Astronesthidae, Chauliodontidae, Stomiatidae, Melanostomiatidae, Idiacanthidae and Malacosteidae, and they all live at various levels in the mesopelagic zone. These fishes also have large jaws and some of the foremost teeth are fanglike. Tine-like teeth on mobile pharyngeal bones at the back of the throat grip and convey sizeable prey to the gullet. Such prey may consist of euphausiids, prawns, squids and fishes, and the gullet, stomach and wall of the abdomen are distensible enough to accommodate large meals. If the predator is stealthy and quick enough food may be a single fish or squid that is at least half the length of its captor. Besides the ventral rows of light organs, the larger and rapacious kinds of stomaitoids have large photophores below and behind the eyes.

The latter cheek lights may be very large and in some species better developed in males. Many species have thousands of small lights all over the skin and in the clear gelatinous tissue that may invest the body. Part from *Chauliodus* spp., nearly all species have a chin barbal bearing beads and inlay of luminous tissue that may be very elaborate. In Chauliodus the second ray of the dorsal fin, is very long, has a luminous tip. Perhaps these are angling devices with a luminous bait. The species of *Chauliodus* and *Stomias*, which are serpentine fishes with a protrusible upper jaw, was slivery-sided with a honey comb scale pattern, framing, at it were, the body lights. The largest species reach lengths of 300 mm or more.

The black of bronze-coloured astronesthids, which range from about 40 to over 300 mm in length, have a conventional fish form and a fin pattern much like that of a smelt. The melanostomiatids are much more diverse in form, ranging from long slim types through fusiform types of short squat types. Female idiacanthids copy, as it were, the black skin and serpentine from that is so common in melanostomiatids, but the males are not only pale-coloured but much the smaller sex. Though also black, the malacosteids have evolved in other directions. The suspension of the jaws is so angled backwards that the gape is very wide.

There is no floor to the mouth cavity and the bare lower jaws with their strong pointed teeth remind one of the lethal mechanisms of a rat-trap. The rat-trap fishes are also unusual in that the cheek photophores produce red light, which is true also of *Pachystomias*, a melanostomiatid. All other luminescent organisms emit blue-green light. Like the smaller kinds of stomiatoids, lantern-fishes are so successful that a sizeable net towed at mesopelagic levels almost everywhere in the ocean is almost bound to catch them. If, as seems very likely they have not colonized the Arctic Sea, then it is improbable that other groups of midwater fishes will have gained a footing.

All lantern-fishes are more of less spindle-shaped and bear much the same fin pattern. The jaws bear rows of small teeth, and though some kinds are larger and wider mouthed than others, there is no marked contrast between small plankton-eaters and larger predators on the micronekton, such as obtains in stomiatoids. While large lantern-fish are naturally able to accommodate somewhat larger food organisms than small species, the diet of all kinds is much as that already stated. The entire size range in the family is from about 25 to 250 mm. While the number and disposition of the lights on the underside of the head and abdomen is more or less standard throughout the group, the constellation of lights that may be seen on a fish in profile seems to be unique in each of the 250 odd species.

When lantern-fishes display these lights, are the individuals of a particular species exchanging recognition signs. There are also differences between the sexes in the development and placing of certain luminous organs. Most of the organs on the head and body look like minute pearlbuttons. The structure and function of lantern-fish lights are considered elsewhere. If the stomiatoids have evolved from smelt-like ancestors, their nearest deep-sea relatives are the argentinoid fishes, which are related to the salmonoid groups.

The argentinoids are small mouthed fishes of diverse body form ranging from short squat types through conventional types to elongated types. Apart from Argentina and its near relatives, which live near the bottom over slope areas, these are fishes of the mesopelagic zone. Their adaptation to this twilight world is evident in their large and sensitive eyes, which are considered elsewhere. Here we need only note that tubular eyes are prevalent in this group. Such eyes, with parallel optical axes that may be directed upward or forward, have also evolved in certain stomiatoids, two families of alepisauroid fishes, giganturoids and the free-living males of the ceratoid angler fish genus, *Linophryne.* In *Opisthoproctus* and certain other deep-sea argentinoids there is a well-developed anal light organ containing luminous bacteria. A unique king of light organ has evolved in fishes of the family Searsidae, which are also mesopelagic and have large and very sensitive eyes. Each organ of a pair, which is set on the shoulder region above the pectoral fin, forms photogenic cells in a darkly pigmented sac that opens backwards through a pore. In *Searsia,* at least, these shoulder organs discharge a bright cloud of luminous particles. Some species have other luminous organs, most of which are set on the underside of the fish.

Apart from *Platyctrotes* and *Platytroctagen,* which have a compressed and keeled body, searsids are fishes of fusiform appearance. The closely related aleopcephalids are much more diverse in form and fin pattern. For instance. *Aulostomatomorpha* has a long tubular snout, *Leptoderma* an elongated body with a long anal fin. *Xenodermichthys* a fairly elongated compressed body and *Alepocephalus* is more or less fusiform with opposed and set-back dorsal and anal fins.

Species of *Alepocephalus,* which may reach more than 0.5 m in length, live near the bottom at depths down to 2,000 metres or more. Some alepocephalids are mesopelagic to bathypelagic. Genera such as *Xenodermichthys* have light organs while others have none. There are over 1,000 species of deep-sea midwater fishes and at least one half are stomiatoids and lantern-fishes. The next most diverse groups are the deep-sea anglers and alepisauroids Lanternfishes and alepisauroids are placed in the same order, but they are very different in functional design and habit. Most of the 100 odd species of alepisauroids belong to the family Paralepididae, which have been called barracudinas because some species remind one of those predatory fishes par excellence, the barracudas. Indeed, alepisauroids, which live from mesopelagic to upper bathypelagic levels, seem to prey largely on other fishes; and they are well equipped to make a living.

Whether spindle-shaped or long-bodies, and there is a marked trend in the group for the evolution of elongated types, the jaws and inner parts of the mouth bear large stabbing teeth or a set of prominent pointed teeth paralepidids. The dentition consists also of many small, pointed teeth that help to grip the prey. Like the predacious stomatoid fishes, alepisauroids have an expandable sac-like stomach, and even in species with body walls of limited distensibility prey up to one-third of the length of the swallower can be accommodated. Nearly all species are non-luminous. Certain of the alepisauroids are among the largest species of midwater fishes. Alepisaurus ferox, the lancet-fish, grows to about 2 metres Anotopterus to about 1 metre and the largest barracudians to about 50 centimetres. The main range is from a few to 30 centimeters.

Though larger than most lantern-fishes, few alesauroids undertake daily vertical migrations that so occupy so many of their relatives. Indeed, the alepisauroids and other sizeable midwater fishes that are adapted to take large meals seem to be non-migrators. Most of the migrators are fishes of modest size with moderate capacities for relatively small food organisms. If, as seems very likely, vertical migrations are fooding sorties, are the migrators 'compelled' upwards to find the best feeding grounds or are there other factors? Are the non-migrators able to stay at one level because of their trophic oppertunism? Giganturoid fishes and the gulper-eel *Saccopharynx* are also adapted to take a wide size range of food organisms. The giganturoids are elongated to long slim fishes with a wide mouth, a fearsome dentition and large forwardly-directed tubular eyes. Their scientific name refers to the very prolonged lower lobe of the tail fin. Like many other mesopelagic fishes, they are silvery.

The species of *Saccopharynx* and the related genus *Europharynx* are very elongated, dark-coloured fishes of lower mesopelagic to bathypelagic levels. Much of their length is taken up by a long rapering tail that is tipped with luminous tissue. The well named *Eurypharynx pelecanoides*, with extremely long jaws that remind one of pelican's bill, grows to a length of about three-quarters of a metre: the species of *Saccopharynx* reach lengths of between 1 and 2 metres. Though *Eurypharynx* has a much larger mouth than *Saccopharynx*, the latter has a more or midable dentition, a much larger stomach and a longer abdominal cavity. Indeed, *Saccopharynx* can be swallow very large fishes but *Eurypharynx* feeds on relatively small prey, such as the young stages of prawns and small fishes. Though the gulper-eels are highly modified fishes the fact that *Eurypharynx*, at least, produces transparent, leaf-like leptocephalus larvae may well mean that the true eels are their nearest relatives.

Midwater eels include the snipe-eels, the bobtailed snipe-eel and *Serrivomer*. The snipe-eels have long, pliable beak-like jaws, set with many minute teeth, and a very elongated body that may reach a length of nearly a metre. They have relatively large, wide-open eyes, where as the bobtailed snipe-eel, like other bathypelagic fishes, has small, reduced eyes. *Cyema*, which has a short tail 'arrow-feathered' with opposed dorsal and fins, may reach a length of about 125 mm. The whale-fishes, which also have small or reduced eyes, are also bathypelagic.

Whale-fishes have large, small-toothed jaws and a considerable capacity, expandable if need be, for copepods and other small zooplankton. One species, *Barbourisa rufa*, which grows to a span of at least 20 centimetres, is one of the few raddish-coloured midwater fishes. The other dark-coloured whale fishes are sometimes tinged with red. Like other midwater species, whale-fishes are widely distributed in the ocean, but they are not commonly caught. Out of some 10,000 fishes representing well over 100 species that were taken off the Canary Islands, Badcock found two specimens and species of whale-fishes. When the whale-fishes have been properly revised they may well prove to be as diverse as the Melamphaidae, a family with which they have some affinity.

Melamphaids, with their round snout, short body, large mid-set dorsal fin and bands of minute teeth in the jaws remind one rather of lantern-fishes. They also have about the same size range. But their thoracic pelvic fins, each with a spin and 6 to 8 jointed rays, are part of a character complex that relates then to the 'lower' spiny-finned fishes. Melamphaids, which are mesopelagic to bathypelagic fishes with brownish to dark colouring, live in all major regions of the ocean, except for the Arctic and Mediterranean Seas. They are without light organs. In one genus there are 'dwarf' and 'grant' species. The former are evidently adapted for life in the impoverished central water masses, the letter for more productive peripheral waters. Fish faunas of neritic and coral-producing seas are dominated by 'higher' spines-finned fishes, but representatives of this great teleost group are in a minority in the deep sea.

Midwater forms include the giant swallowers and the trichiuroid fishes, predatory types with marked piscivorous ways. *Chiasmodon niger* can swallow fishes larger then itself. The trichiuroids are diverse in form, ranging from elongated types, such as *Paradiplospinus*, *Gempylus*, the snake mackerele and *Aphanopus*, the black scabbard fish, to those with a fusiform body, e.g. *Ruvettus*, the oil fish and *Nealotus*. Trichiaroids, which are somewhat reminiscent of alepisauroid fishes in form, fearsome dentition and sleek appearance, are represented also by a number of large species. The snake mackerel grows mackerel grows to about 1 metre, the black scabbard fish to 1.25 metres and the oil fish to over 1.5 metres. There is a deep line fishery for the second species off Madeira. The depth of transition from the twilight of the mesopelagic zone to the sunless waters of the vast bathypelagic zone is near 1,000 metres. Dark-coloured species of Cyclothone, whale-fishes, gulper-eels and the deep-sea angler-fishes, which related to the angler-fishes or monkfishes and frog fishes, form most of the bathypelagic fish fauna.

Angler-fishes bear a rod and bait on top of the head, but in the ceratoids it is developed only in the females. In all but one form the bait is luminous and differs in structure from species to species. In each species the individuality of the lure may be seen in its shape, photogenic complex and dressing with dermal appendages, but there is no evidence yet that particular lures attract particular kinds of food organisms. It seems more likely that female angler-fishes, which have a wide, well armed mouth and distensible accommodation, are trophic opportunists, able to take a wide size range of prey. Perhaps the males are prawn to proper partner after receiving the 'right' sign stimuli form her lure. But the males of a few species have tiny eyes, though their olfactory organs, like those of most male ceratioids are highly developed. Perhaps the females with keen-nosed maters lay scent trails during the breeding season. The males are much the smaller sex and their dwarfing is but one aspects of a complex and trenchant sexual dimorphism. After metamorphosis the males are about 10-40 mm in length, and there is little subsequent growth. They are active more or less fusiform fishes but in nearly all species the females are tubular or pear-shaped.

In certain ceratoids the male becomes parasitic on the females and in such forms she is from three to more than ten times the length of her mate. The free living males of *Himantolopdus*, which is one of the ceratoids that seem to have other mating arrangements grow to about one-thirteenth of the length of the largest known female. Though their means of reproduction are unusual, the life-history patterns of deep-sea angler-fishes are much like those of most other mid-water fishes, whether mesopelagic or bathypelagic. Their eggs, which must be very numerous but have yet to be found, are probably shed in the depths and develop as they float upwards. During metamorphosis to adult form there is descent to the levels of the adult living space. Here a few of the young, fishes will eventually reach maturity.

ZOOPLANKTON

This comprises representatives of almost every major group of animals, either as adults or as larvae. Of the 135000 species of bottom-dwelling invertebrates, 1,00,000 have a free-swimming larval phase which lasts on average about two to four weeks. Just as there were single-celled forms of phytoplankton, so there are single-celled members of the zooplankton. These are not capable of producing their own food but are consumers relying upon ingestion of other organisms, usually smaller phytoplankton and bacteria. Some Protozoa produce elaborate shells or tests and include the radiolarians and foraminiferans. Globigerina is an example of the latter and enormous numbers occur in warmer seas and when they die, accumulate on the ocean floor. This forms Globigerina ooze, a sticky glutinous sediment covering large tracts of ocean bed e.g., in the Tasman Sea Coelenterates. Both holoplanktonic and meroplanktonic species exist.

In the class Hydrozoa there are two main phases to the life history-bottom dwelling colonies of hydroids which give rise to meroplanktonic reproductive bells such as *Bougainvillea ramosa*. The common jellyfish Aurelia aurita which has a world-wide distribution, does not have the usual sessile hydroid stage. Larvae produced from its reproductive bell, settle and then bud off new larvae known as ephyrae, rather like a pile of saucers.

Ctenophores

These include the common sea gooseberry; they are holoplanktonic and unlike coelenterates have no stinging, cells. They swim by beating eight rows of 'comb-plates' Larval herrings form an important part of the diet of the sea gooseberry *Plaurobrachia.*

Chaetognaths

Otherwise known as arrow worms these are holoplanktonic and transparent, torpedo-shaped, active predators. Various species of *Sagitta* are found in British waters and were used by Russell in 1939 as biological indicators of different water masses. Certain species such as *Eukrabnia* hamata tend to be found at shallow depths in high latitudes and greater depths in tropical regions to compensate for the rise in temperature of low latitude surface waters but departures from these patterns of distribution in the water column are not uncommon.

Polychaetes

A few polychaetes are permanently planktonic as adults and include *Tomopteris belgolandica* which may reach 35 mm in length and is found in the North Sea and off Norway. Each segments of the body has lateral extensions with two branches known as parapodia, much enlarged and flattered in these planktonic forms. Many other polychaetes are benthic and only become planktonic during the breeding season. Sexual forms are known as epitokes which have special adaptations for swimming such as the enlargement of eyes, muscular parapodia and bundles of bristles or Satae. The activities of the Pecific Palol worm *Eunice viridis* off Samoa are now legendary. Every October and November on two days at the last quarter of the moon there is a mass swarming of epitokous tail-ends of Palolo worms at the surface of the sea. Unmodified head-ends retreat into crevices among the corals on the bottom. Native fishermen scoop up the swarming worms by the basketful and provide a once-a-year delicacy. Some of the syllid polychaete epitokes such as *Autolytus* undergo an elaborate mating process with the male encircling the female with sperm threads prior to the female shedding eggs into a ventral egg sac.

Crustacean

These are by far the most numerous species, comprising about 90 percent by weight of all zooplankton. Holoplanktonic copepods occur in vast numbers and the calanoids predominate in most seas. *Calanus finmarchicus* comprises the food of the herring in the North Sea. Copepods have adopted various strategies to avoid being eaten. There is no shelter in the open sea, hence in the surface waters most large copepods are transparent. In deep water copepods are often red in colour but are able to hide under cover of permanent darkness. In middle depths the copepods migrate up to the surface at night where they feed in the phytoplankton-rich waters. At dawn they migrate down to avoid predators; the larger they are the deeper they need to go in order to be safe.

Species such as the amphipod *Gammarus oceanicus* are found in the open sea in the Arctic were salinity is reduced from the effect of melting ice. Further south they are found only in coastal waters where there is some dilution from rivers. Many other crustaceans are meroplanktonic including the cirripedes or barnacles. They begin life as a free-swimming nauplius larva with two easily identifiable horns projecting laterally at the front of the carapace. After several moults a ypris larva is produced with a bivalved shell. This stages settles and the familiar calcareous plates of the adults are secreted around the transformed animal, now filter feeding by means of the legs extended through the opening, guarded by smaller plates.

The malacostracan crustaceans include both holoplanktonic forms with a dispersal phase in their life histories. Euphausiids are widely distributed throughout the oceans and seas of the world but perhaps the best known is 'krill' *Euphausia ouperba* of the Southern Ocean. Females spawn 2000 to 5000 eggs which then sink 800 to 1000 m before hatching into nauplius larvae with three pairs of swimming appendages. These larvae moult and graduall acquire more swimming legs and the characters of the adults. The rise to the surface by larval stages of takes a few weeks and this 'developmental ascent may be aided byupwelling currents. The decapods include may meroplanktonic forms.

Penaeid shrimps such as the American Pink shrimp *Penaeus duorarum* fished in the Western Atlantic have a long life cycle. There are several phases beginning with the nauplius, changing to a protozoea, followed by a mysis, a post larva and finally the juvenile which has most of the specific characters of the adult.

Caridean prawns which predominate in European waters pass much of their early larval development in the egg membranes hatching at the zoea phase, several moults then occur prior to metamorphosis. The stage 1 palaemonid zoeal larva has sessile compound eyes with apparently hexagonal facets, three pairs of biramous thoracic swimming appendages, a solid triangular telson and swims upside down, tail first. After metamorphosis the post larva has developed stalked compound eyes with square facets, lost the outer swimming branches of the thoracic limbs, developed abdominal pleopods which are used for swimming, gained the typical tail fan with lateral uropods and swims the right way up, heat first. Crabs often have short life cycles with only two phases: zoea and megalopa.

Molluscs

Some complete their development of the crawling stage inside the egg, but most hatch as a free-swelling planktonic larva, which, after this dispersive phase, usually become benthic The first larval stage in the common European limpet *Patella vulgata* is a trochophore less than 0.2 mm across. Rows of cilia forming prototroch beat and rotate through the water. Two days after fertilization the larva is transformed into a veliger with a shell and a typical molluscan foot. While a planktonic veliger the limpet undergoes torsion. The first 90^0 twist is brought about by muscle contraction. After 30 hours full 180^0 torsion is complete. After 3.5 to days pelagic life ends, the velum, disappears and the larval shell is lost. The edible periwinkle *Littorina littorea* hatches as a veliger with a 1.5 whorled shell. The trochophore stage is passed in the egg. The pelagic phase lasts for up two months and hence gives great possibilities for dispersion. Nanoplankton is wafted into the mouth by the beating of the velum cilia during the planktonic phase.

Echinoderms

Adults are radially symmetrical but their larvae show bilateral symmetry and are usually planktonic. Eggs are shed into the sea and fertilization is external. Normally there is no brood protection but there are exceptions, particularly in polar seas. A common larval form links the groups of echinoderms and is known as the dipleurula or 'little sides'. In the early embryo the whole surface is covered with cilia forming a closed loop. In sea cucumber larvae the band becomes sinuous. In starfish larvae the ciliated band divides into two and again becomes sinuous. Sides of the larvae are drawn out into long process, the three are the anterior end bearing adhesive discs. The starfish forms inside and after about two months breaks free. In sea urchins the larva is known as a pluteus and in brittlestars an ophiopluteus.

Fish

The eggs and larvae of fish are most often planktonic. But even eggs that sink or are attached to the bottom of, for example, gobies, wrasse and herring may give rise to planktonic larvae. Yolk is utilized initially until the mouth and anus become functional. Fish larvae are often quite unlike the adult and identification is difficult. Specific patterns of pigment cells may be used before the more reliable meristic characters of ossified by skeleton in the post larva has developed.

Reproductive Strategy

In late summer and early autumn in some temperate waters such as the southern North Sea the meroplankton. Often greatly out number the holoplankton. The reason for this dominance is that they are often produced in vast numbers. This kind of reproductive strategy is adopted by 'r' selected species in which large numbers of small eggs with little or no yolk are produced. The larvae which hatch from these eggs are planktonic and often have a long pelagic life and of necessity start feeding immediately. If there is an abundance of other small planktonic organisms then these opportunists take advantage of the food supply and many

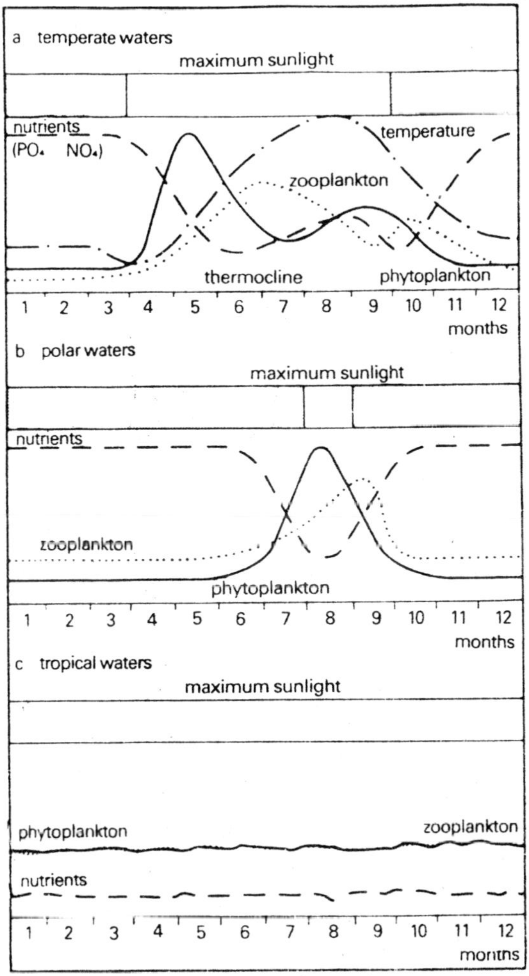

Fig. 8.21. Seasonal abundance of phyto- and zooplankton.

survive, some settling in new areas, the larval period allowing good dispersion. If for any reason small planktonic food organisms are in short supply, then many of the larvae die. Within these very large broods there is considerable potential for genetic variation. If some of the larvae meet with unusual conditions they might still be able to survive. Many invertebrates such as some polychaetes, molluscs, echinoderms and some crustaceans produce many millions of eggs in a breeding season. Mortality is often very high. But if more than two individuals survive, the species could soon build up large populations, dominating and possibly excluding other species from the area. This rarely happens because physical and biological controls such as competition, food supply, availability of places of hide usually prevent a species from over-running an area. These natural controls tend to maintain balanced and stable populations. The other kind of reproductive strategy adopted to 'K' selected species involves the production of small numbers of large eggs rich in yolk. They either develop directly into miniature adults or have a very short free-swimming larval existence. In these forms the larvae do not need to seek food immediately.

There is a complete gradation between the two extremes of r' and 'K' selection. In 'K' selected species there is a greater certainty that the small numbers which are produced are well provided for and will survive. Thus reproductive effort is either directed at large numbers of eggs poorly supplied with yolk or small numbers with more yolk and direct development. Abnormal sea temperatures may retare development. Larvae actively seek out the preferred kind of habitat, but larvae will delay metamorphosis, e.g. shrimps larvae achieve this by inserting extra moults, if conditions are not favourable. It has been suggested that certain 'giant' caridean larvae have been carried far from their normal habitat and have continued to grow without metamorphosing into the adult form. However, all these delays result in the larva being exposed for longer to attack from active predators or even filter feeding molluscs. Up to 10000 larvae of a variety of marine invertebrates may be tatrapped in 24 h at the height of the summer season by a singlemussel. Even settling may be hazardous where coelenterates lie in wait armed with batteries of stinging nematocysts.

Seasonal Variation

In temperate waters increasing hours of sunlight and to a lesser extent the spring rise in temperature herald a phytoplankton bloom.

This is followed closely by a peak of zooplankton grazing on the phytoplankton. This grazing and a lack of nutrients caused by a stable thermocline–a sharp boundary between warm and cold water –causes a crash in phytoplankton populations. In autumn the thermocline breaks down under the influence of equinoctial gales stirring up the water column and bringing to the surface nutrients that had been dissolved in water too deep and therefore too dark for plant growth. This results in a second smaller phytoplankton bloom. In the tropics the plankton population is much more stable throughout the year. In polar seas there is no phytoplankton production during the six-month period of permanent or semi-darkness. There is also no thermocline, so nutrients are not limiting, but phytoplankton production is confined to a brief period at the height of summer followed quickly by a single zooplankton peak.

9

BOTTOM DWELLING ORGANISMS

Biological communities consists of groups of populations of species living together in a particular environment. Living organisms and their non-living environment are always closely interrelated and interact with one another. The benthic or bottom-dwelling community lives on and in the floor of the seas and oceans, and ranges from kelp forests and coral reefs in shallow seas to the unusual fauna associated with soft sediments in the abyssal zone.

Collecting and Recording Gear

In the early years of this century Petersen designed and constructed a bottom sampler. He used it on soft sediments in shallow water and although the grab was effective, the samples were small, only 0.1 or 0.2 m^2. The relative sparseness of many benthic organisms meant that the number of specimens per sample was small. Single samples therefore gave an unreliable indication of the abundance and dispersion of the fauna sampled. The depth to which Petersen's grab sampled depended on the nature of the substrate. In

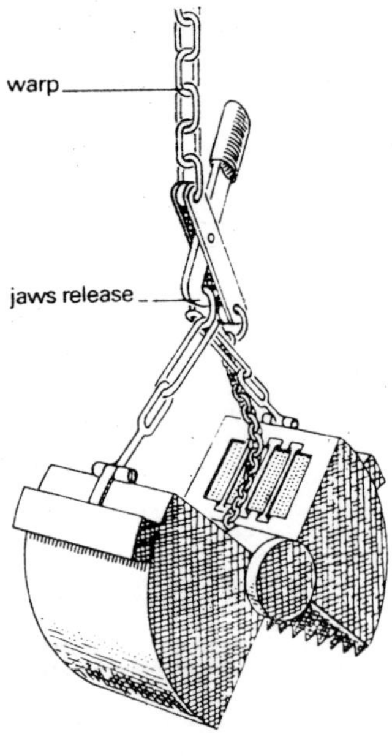

Fig. 9.1. Petersen's grab.

soft mud penetration was good, but some of the deeper burrowing forms were not sampled at all. Various workers have modified Petersen's basic grab in order to overcome the problem of shallow bites on harder grounds One examples is the Van Veen grab which has long scissor-like arms, closing the grab with a lever action.

Smith and MacIntyre devised a grab using the Van Veen lever system but mounted in a frame with two trigger plates. The jaws do not fire until the grab has settled on the bottom. The jaws are then operated by strong springs that thrust the grab into the substrate, Another variation is the Holme double scoop grab which

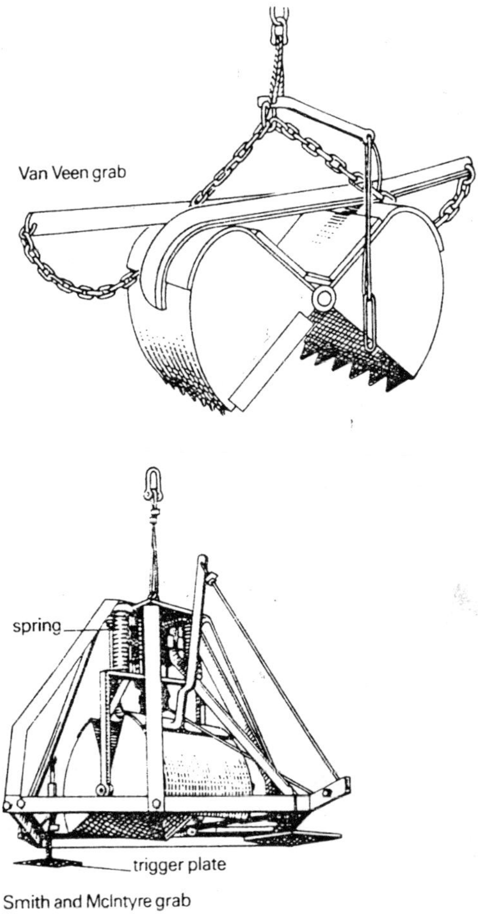

Fig. 9.2. Infaunal samplers: Van Veen grab and Smith and McIntyre grab.

depends on a weighted frame to achieve a deep sample. These grabs all give reasonably quantitative estimates of numbers per unit area. Grabs sample relatively small areas so there is a need to take several samples in order to obtain a representative selection of all species present. This operates on a law of diminishing returns– after an initial worthwhile increase in the numbers of species taken, subsequent grab samples produce fewer species that have not been collected already. Another grab that is very effective at penetrating hard, compacted substrates is known as the Orange Peel grab

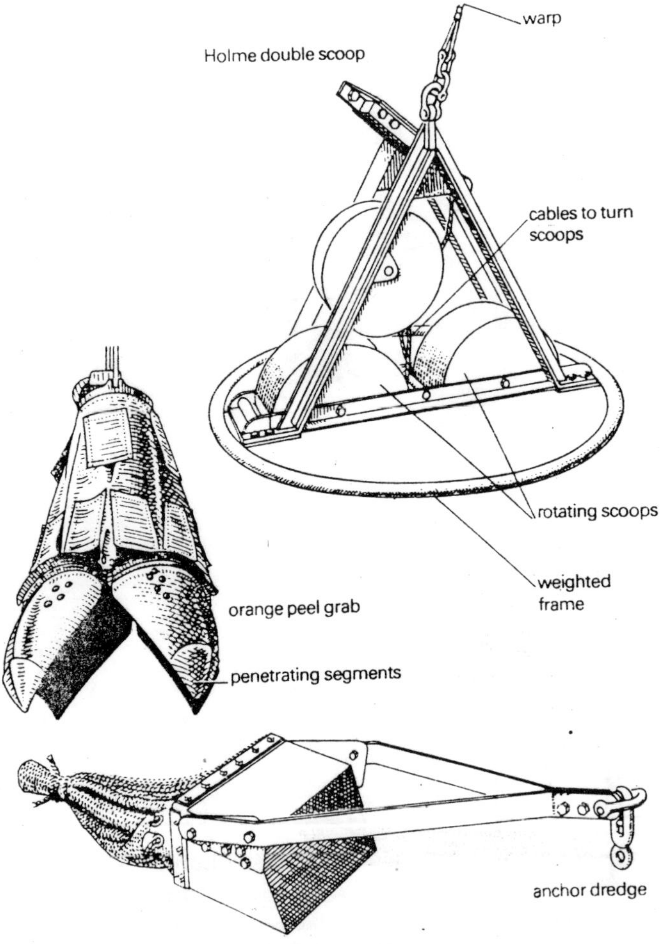

Fig. 9.3. Infaunal samplers: Holme double scoop, Orange peel grab and anchor dredge.

because it consists of several 'segments'. It suffers from the disadvantage that it is impossible to estimate the area covered by the grab and the results are therefore qualitative rather than quantitative. The anchor dredge is a more useful means of taking a semi-quantitative sample. All of these sample the infauna or those animals living within the sediments.

The other group of benthic organisms the epibenthos is collected most abundantly by dredges and trawls. Animals live on the surface of hard and soft bottom substrates. There are many variations on the basic trawl for collecting epibenthos including a beam trawl which is strong enough to be used on rough substrates. On very rugged substrates an Agassiz trawl may be used which has the advantage over a beam trawl in that it will continue to sample either way up. If the going becomes really rough then a scallop dredge is used which looks rather like a trawl except the mesh is made of interlocking steel rings. The collecting gears described so far all operate remotely and it is often, difficult to judge just how effective they are at collecting specimens because except in very shallow, clear water it is not possible usually to monitor progress. Diving has enabled both direct observation of organisms *in situ* and closely controlled hand collecting.

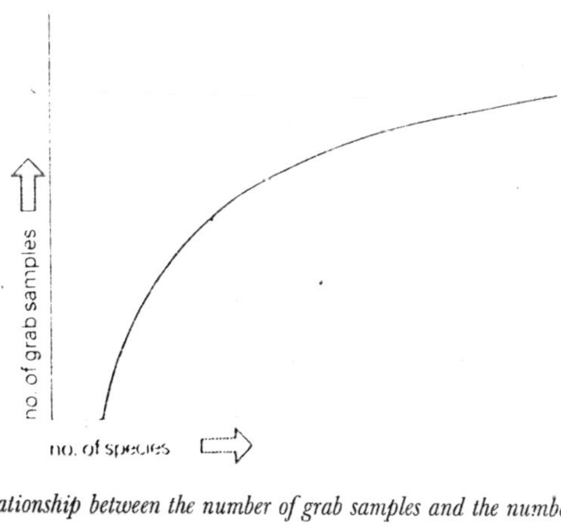

Fig. 9.4. Relationship between the number of grab samples and the number of species collected.

In 1943, Jacques Cousteau collaborated with Emile Gagnan, an engineer working on a gas regulator used in war-time Paris taxis running on coal gas. Together they designed a demand valve and the subsequent development of the Self-Contained Underwater Breathing Apparatus gave great impetus to this direct approach to marine biology. Divers using aqua-lungs are able to make *unique contributions* to biological recording schemes and collecting data for this purpose has become a regular/feature of many club dives. Diving is a particularly valuable method of obtaining reliable information from hard irregular substrates where it would be difficult to operate even a steel dredge. Depths of 40 m may be achieved comfortably by competent wet-suit divers using compressed air and dives to 60 m are technically feasible, with proportionately longer stops for decompression on the way up. On deep dives this means that the amount of useful working time on the bottom is limited. Longer dives are possible in more cumbersome hard helmet dry suits or by saturation diving.

The latter method exploits the fact that once a diver has been compressed at a given depth for about 10 hours his body tissues absorb no more inert gas. The diver is able to stay at that depth indefinitely without needing extra decompression time at the end of the dive. Teams can work in deep water for days without returning to atmospheric pressure by being transported from the bottom to the surface in a compressed diving bell and then transferring to a deck decompression chamber where they eat and sleep, still at the same pressure as the dive site on the bottom.

Spectacularly deep dives in manned submersibles such as the American *Alvin* provide a means of direct monitoring and collecting at 2500 m and more. Electronic flash photography is used increasingly by divers. This method of recording is used also in conjunction with sophisticated deep water epibenthic trawls. Acoustic beacons relay information back to the ship to ensure that the trawl is in contact with the bottom. An angled camera, attached to the frame of the trawl is often used to take shots ahead of the net at regular intervals. These photographic data of the benthos are necessarily restricted to the epifauna (and flora in shallower water) but they do give clear information on abundance and dispersion. This visual record is often at variance with what comes up in the trawl.

There is almost always less in the trawl than one might expect from the photographs then known distance covered as measured by an odometer. Another method of visually recording events under water is by closed-circuit television. Resolution is often not as good as photographs but it does enable continuous observations of animals in their natural habitat to be made. But all these methods are

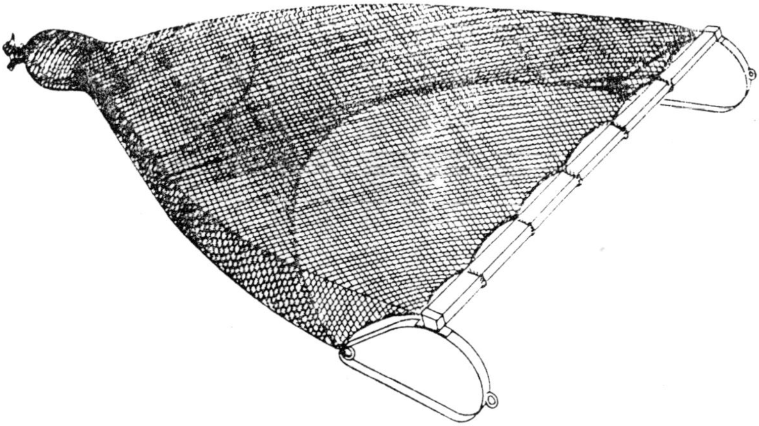

Fig. 9.5. Beam trawl.

expensive. However, in order to obtain a representative sample of the benthos ideally a trawl with a camera attached should be used in combination with a deep sampling quantitative grab. Even with the array of equipment there are uncxpected difficulties; as a crew becomes more adept at handling gear, so the number of species caught increases. This inbuilt bias in a series of a samples will be difficult to estimate.

Physical Factors

Temperature

Groups of a marine animals may be divided up into regions bounded by sea temperature isotherms. Animals occurring at great depth are likely to be stenothermic (able to tolerate only a narrow range of temperature) and hence would be unlikely to penetrate into shallow water where temperature is more variable. The 20°C isotherms to the north and south of the equator effectively delimit the distribution of coral reef formation.

Salinity

Average salinity near the surface is reduced in high lattitudes from the dilution effect of ice melting. The relatively calm Sargasso Sea area at the centre of the North Atlantic gyre has a consistently high salinity of over 37 parts per thousand. Salinity becomes lowered in tropical regions only where there is a significant dilution effect from rivers such as the Amazon and Niger.

Light

Visual feeders are obviously directly affected, but light also indirectly influences all benthos in that most of the food web is dependent upon plants fixing the energy of sunlight through photosynthesis. One of the difficulties in examining the organization of benthic communities is in finding out what powers them and keeps them going. Below the photic zone these is very little primary production with the exception of some autotrophic bacteria. Nevertheless there are vast areas of underwater communities in complete or semi-darkness surviving without any direct energy input from the sun.

In the photic zone most of the plants are algae, either usually fixed macroalgae forming dense kelp forests in cool temperate waters or unicellular forms, mostly planktonic but also forming vast populations of symbionts in the tissues of coelenterates and other animals, particularly in coral areas. There are a few flowering plants in the shallow water zone such as the eel grass *Zostera.* When Petersen was taking grab samples earlier their century he estimated that in Danish waters alone there were 24 million tonnes of eel grass growing in dense 'meadows' in the shallow waters. He assumed that the dead and dying vegetation from these meadows became detached and was carried by currents and tides into deeper waters and in the form of detritus undergoing microbial decomposition thus provided the source of energy for his marine benthic communities. This idea seemed sensible and went unchallenged until 1931 when there was an eel grass epidemic disease caused by a fungus. Vast areas disappeared in a few years. But the benthic communities continued apparently unchanged. It seems likely that the phytoplankton that forms the basis of the plankton food web, also forms the source of energy for the benthos as well, in the form of a continual rain of detritus from above.

Exposure and Desiccation

Intertidal animals of necessity have become adapted to continual wetting and drying out but desiccation is a powerful barrier to the shore-ward extension of sublittoral species.

Oxygen

In certain fine sediments anaerobic conditions may develop and effectively exclude many species requiring a good supply of oxygen.

Many of the meiobenthic forms thrive in these deoxygenated conditions.

Currents and Tides

These will affect the dispersal of organisms especially those benthic forms that have larval stages forming part of the meroplankton.

Sediment

When a grab sample lands on deck the animals are sorted from the sediment by sieving, preserved and usually identified later in the laboratory. Before this sieving takes place a small sample is usually kept intact for sediment analysis. After gently drying the sample it is analysed using a stack of sieves ranging from coarse at the top to very fine at the bottom. The proportion of different grades of sediment in the sample are then worked out. Bottom sediments are often very rich in organic matter-up to 10 percent-and this again is easily determined on an approximate basis in the laboratory. The sample is first dried gently to drive off the moisture content, weighed, heated fiercely to burn off organic matter, cooled and reweighed. However, carbonates are

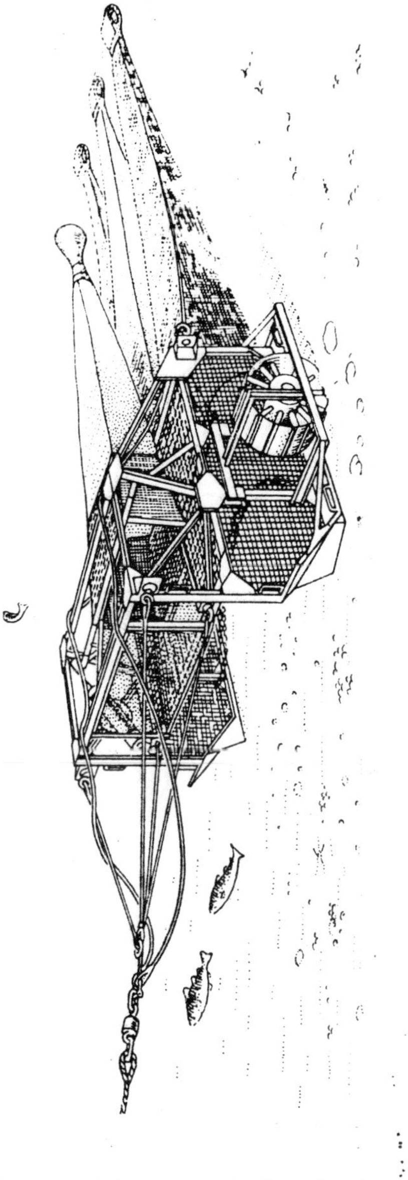

Fig. 9.6. Deep water trawl sampling the epibenthos.

inadvertently included in the weight loss of the sample. A more accurate method of estimating organic matter which excluded carbonates involved heating with chromic acid, followed by volumetric analysis. This has now been superseded by automated carbon and nitrogen analysers.

Biological Factors

Many physical factors exert their effect through the biological properties of the organisms. Breeding seasons are often triggered by a particular temperature. However, the relationship is usually far from direct. Other factors may delay the onset of breeding even when the threshold temperature has been exceeded. Some factors are independent of the physical environment, for example, relationships with other organisms such as a predator and its prey and also parasitism. Competition for space can limit the number of particular species, These biological factors are much more difficult to estimate than physical factors.

Benthic Organisms–Infauna of Soft Sediments

Polychaetes

A variety of worms present in large numbers live in burrows. In shallow water on mud and sand the ubiquitous lugworm *Arenicola marina* may be found. *Nephtys caeca* prefers exposed sandy coasts.

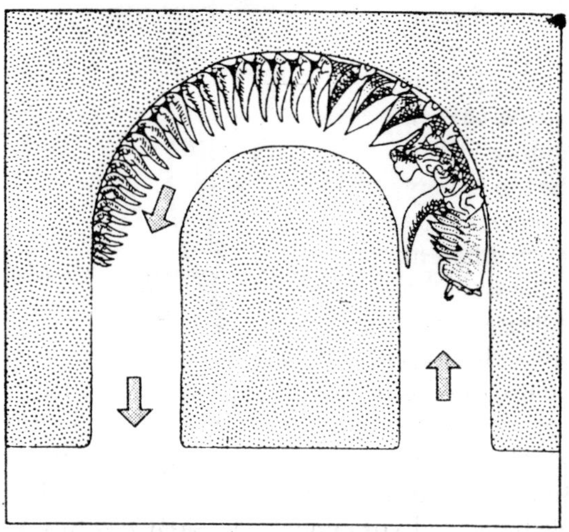

Fig. 9.7. Polychaete Chaetopterus in parchment-like tube. Arrows show direction of water current through tube.

The highly differentiated tube-dwelling *Chaetopterus variopedatus* has the middle segments of its body specialized into three flap-like paddles that draw water and food through the parchment-like tube.

Molluscs

There are many bivalved molluscs that burrow into soft substrates using hydrostatic pressure in the foot to punch a way through the sediment. These bivalves have inhalant and exhalent siphons used for drawing water over gills of extract food particles. Molluscs were used by Petersen in naming benthic communities including *Venus gallina* of the open coast of the Kattegat and the North Sea, found in association with *Tellina fabula* and the heart urchin *Echinocardium cordatum*. Another *Venus gallina* community is found in deeper water in the North Sea in association with different species of molluscs and echinoderms.

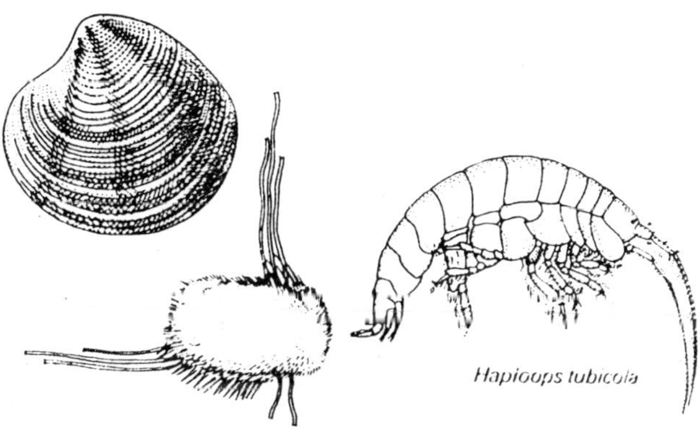

Fig. 9.8. Infauna. Venus gallina, a suspension feeding bivalve mollusc. Echinocardium cordatum, a deposit feeding heart urchin. Haploops, a tube-dwelling amphipod.

Crustaceans

These form a prominent part of the benthic infauna. Adult scampi *Nephrops norvegious* burrow in mud in depths from 5 to 70 m and cat polychaete worms, small fish and other smaller crustaceans. By far the most dominant groups of benthic crustaceans are small cumaceans and amphipods. Petersen designated an amphipod dominated community–*Haploops tubicola* in association with bivalves. Infaunal amphipods have been used as indicator species for communities in the shallow-water sand habitat of the Irish Sea.

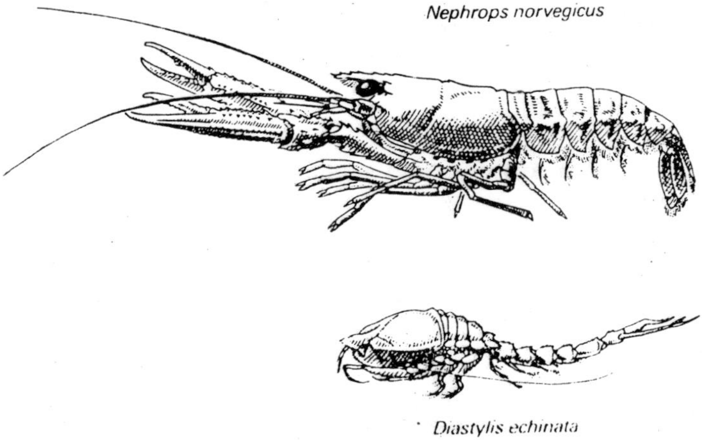

Fig. 9.9. Infauna. Nephrops norvegicus, a scavenging decapod crustacean. Diastylis echinata a cumacean grazing on sand grains.

Series of samples were taken using a 0.1 m^2 Van Veen grab. Analysis of the amphipods present showed that abundance and the number of species was greater in the west.

One genus–*Bathyporeia*–dominated, but zones dominated by individual species of *Bathyporeia* could be identified also. These could by correlated with physical variables. Sea water temperature

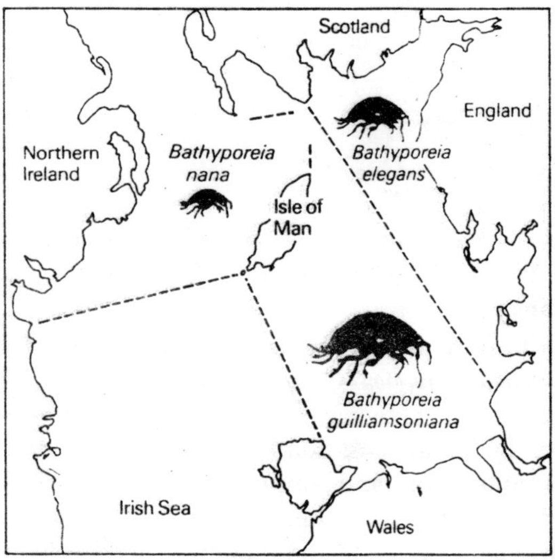

Fig. 9.10. Amphipod zones in the shallow water sand of the Irish Sea.

varies in the deeper western Irish Sea from 7 to 13.5°C only, whereas in the shallower eastern areas it ranges from 5 to 16°C. Research has shown that pollution may depress species diversity in the east and in Liverpool Bay in particular. Sediment size may also be a factor influencing the distribution of amphipod communities.

Bathyporeia guilliams niana is the largest species the genus and is able to inhabit coarser sediments. Sand particles are manipulated by the mouthparts and organic matter scraped off and passed to the mouth.

Bathyporeia nana, dominant in the west, is effectively excluded from coarser sediments because of its small size. Groups of amphipod species often occur together and are good indicators of the shallow-water sand community.

Echinoderms

The heart urchin *Echinocardium* and the little heart urchin *Echinocyamus* live buried in sand. A starfish *Astropecten* also inhabits sand and feeds on small bivalves. The sea cucumber *Holothuria* prefers muddy sediments and captures prey by entwining them with sticky mucous threads. An abyssal sea cucumber.

Benthic Organism–Epibenthos

Organisms living on bottom surfaces (epibenthos) differ markedly depending on the nature of the substrate.

Hard Substrates

In shallow cool temperate waters kelp forests dominate and the hold-fasts, stipes and fronds provide suitable surface for many fixed and mobile invertebrates to live. Fixed forms associated with kelp and rock surfaces include many coelenterates such as hydroids, anemones and corals, sea squirts. The mobile species include some of the larger decapod crustaceans such as crabs but also many polychaete worms and echinoderms such as the sea urchin *Echinus esculentus.* Under rocks, pebbles and seaweeds large concentrations of brittlestars may occur, e.g. *Amphipholis squamata* which is recorded from lower shore down to 250 m. One major benthic community in tropical regions that produces its own hard substrate is that of corals and coral reefs.

The reefs and formed from the calcareous skeletons of the colonial coelenterates. Distribution is limited to shallow seas where the temperature does not fall below 18.5°C; coral reefs are one of

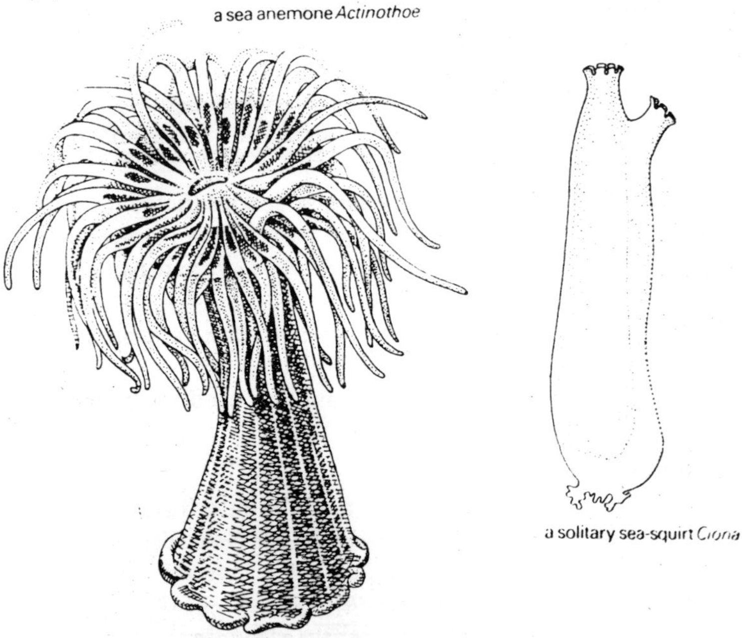

Fig. 9.11. A sea anemone–Actinothoe and a solitary sea-squirt–Ciona.

the most biologically productive of all communities. Water flow over reefs and the efficient re-cycling of nutrients are the factors largely responsible for the high productivity. Although the polyps are animals, the reef is a virtually self-contained ecosystem with its own food web. Algal zooanthellae live in the tissues of the polyps and other algae live in the hard calcareous part of the reef. These algae need to photosynthesize and hence restrict live coral to 90 m depth at most. This symbiotic relationship between the coral and algae is thought to have contributed to the success of the reef ecosystem. The relationship appears to allow nitrogen and phosphorus to be recycled. Hence the typically low levels of these nutrients in tropical waters are no longer a limiting factor in the reef ecosystem.

The coral polyps also contribute to reef productivity by capturing zooplankton and phytoplankton as the waves pound in. Reefs appears as 'oases' with a gross production greater than 10 g per square metre per day surrounded by open ocean 'desert' of very low productivity. Some corals do not have symbiotic zooanthellae and are capable of living at greater depths, but they

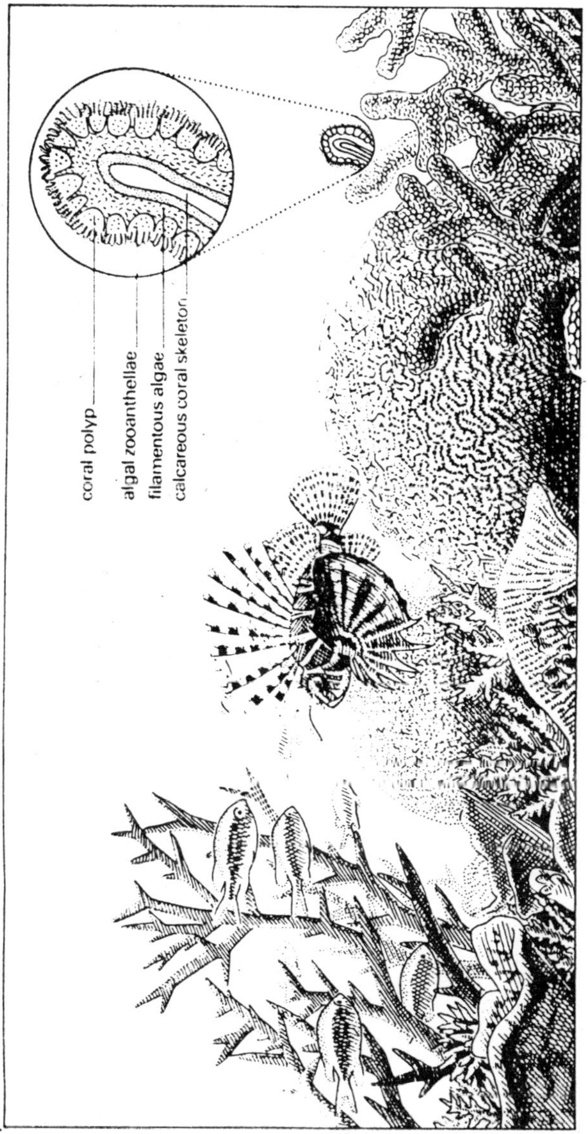

Fig. 9.12. Life in a coral reef lagoon.

are generally solitary and not reef-forming. Darwin described three types barrier reefs found along continents, fringing reefs around islands and atolls, horseshoe-shaped reefs centred around a lagoon. Living reefs that are located in mid-ocean are usually situated on the top of a volcanic cone.

In the some atolls there is a great depth of dead coral and this is used as evidence of sea-bed sinking or rising sea levels. The coral has been able to keep pace with the change by growing upward, but any below the photic zone dies. At night the coral polyps extend their tentacles to capture plankton rich in scarce nutrients such as phosphorus. Vigorous growth occurs in regions where there is good current flow, especially through channels in the reef. On the inner edge of the lagoon silting may depress growth while surf breaking on the outer edge may damage new coral growth; encrusting coralline algae thrive in these conditions of pounding surf. Reefs attract a dazzling array of other inhabitants ranging from conspicuous shoals of fish to anemones, crustaceans and sponges while nearby, the top predators-sharks and moray eels-lie in wait for the unwary.

Soft Substrates

In deeper waters there is a tendency towards a uniform muddy bottom. Around the poles this is principally composed of diatomaceous ooze, mainly of silica whereas Globigerina ooze dominates in temperature and tropical oceanic regions and is mainly calcareous. Some of the more unusual epibenthic fauna including Venus's Flower Basket Euplectella, a deep-sea glass sponge living between 450 and 900 m. A commensal relationship has developed with a species of shrimp. A young male and female shrimp enter the hollow cylinder formed by the fusing of supporting spicules into a lattice. The end plate sealing off the cylinder effectively imprisons the shrimps which feed on plankton carried on water currents through the lattice. Coelenterates related to the corals known as a sea pens Pennatula have a large primary polyp anchored in the substrate; feeding polyps occur as branches off the main stem. Priapulid 'worms' are predaceous and capture slow-moving invertebrates by everting the invaginated mouth region armed with rows of spines. A deep water asellote isopod has elongated legs which enables it to move easily over the surface of soft ooze. Other epibenthic fauna photographed from a deep-water trawl.

Meiobenthos

This important group of animals includes all those interstitial forms living in the spaces between sand grains. The type of animals range from rotifers, nematodes, harpacticoid copepods, kinorhynches, turbellarian worms through to minute sea cucumbers and sea squirts. Meiofaunal populations of 4 million m^{-1} are not uncommon in

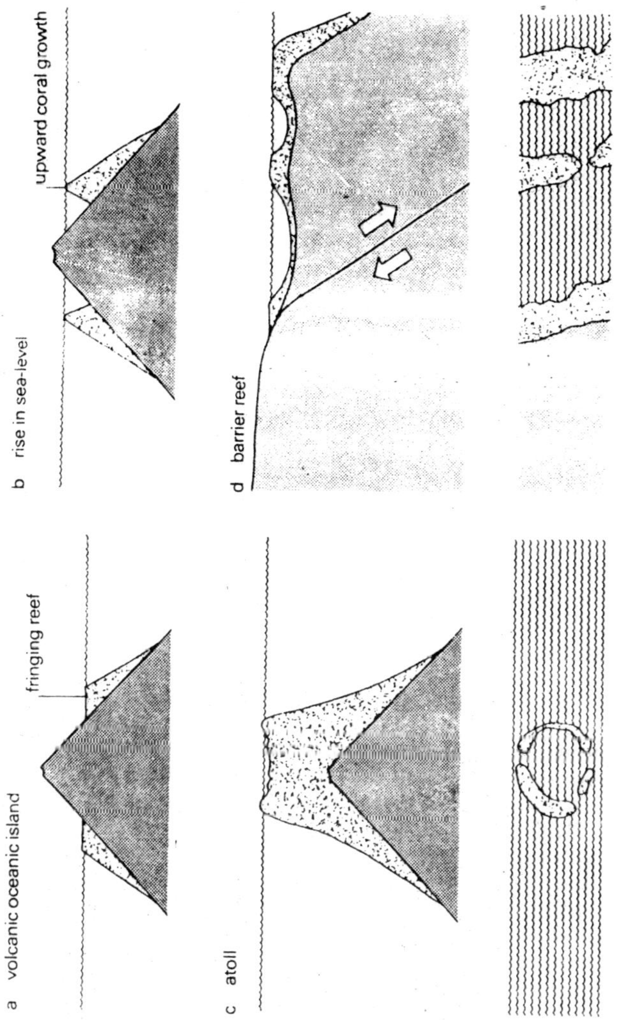

Fig. 9.13. Darwin's theory of the formation of coral reefs. (a)-(c) Stages in the development of mid-oceanic fringing reefs and atolls showing the effect of a rise in sea-level (or of sea-bed sinking). (d) Continental barrier reef and shelf movements. Aerial views of atoll (bottom left), barrier reef (bottom right).

shallow sea and their biomass average 10 to 20 percent that of the macrofauna. Their importance in benthic ecology and in benthic food webs in particular is only beginning to be appreciated. Many microfaunal groups such as bacteria, protozoa and meiofaunal nematodes have representatives capable of living in deoxygenated or anaerobic conditions.

It has been suggested that the origin of life on earth may have taken place in Precambrian deoxygenated estuarine muds and that

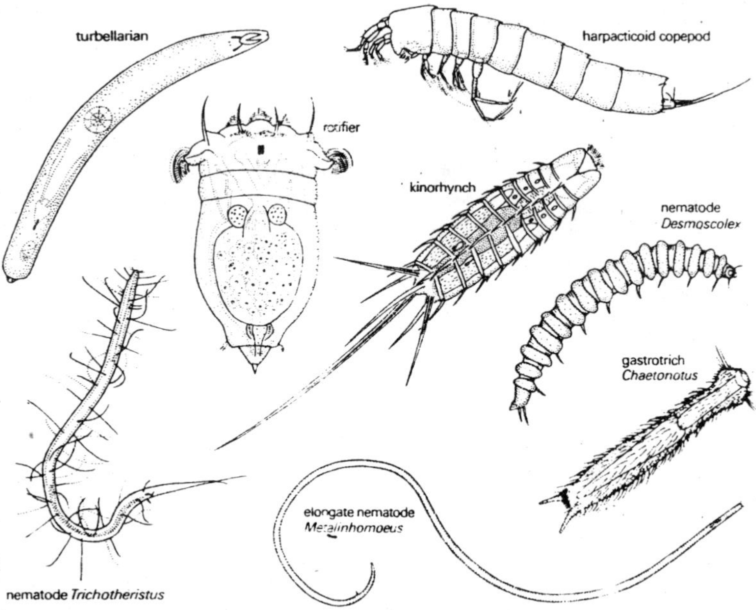

Fig. 9.14. Meiobenthos.

early metazoan forms were benthic anaerobes absorbing nutrients through their outer epidermal layer. Further, it is claimed that anaerobic metabolic pathways were able to switch to aerobic respiration when the oxygen levels in the atmosphere had built up following the development of photosynthesis in autotrophs.

Community Analysis

The idea of classing marine benthos into communities was largely established by Petersen in 1913 when he divided up the fauna of Danish waters into communities based on a survey of the animal caught in his grab sampler. He found that there were certain common species, especially of molluscs and echinoderms, that would act as indicator species. From even a small sample it was possible to predict the kind of community from which the sample had been taken by looking for these indicator species. He usually chose a pair of species to name a community such as the Echinocardium-filiformis community typical of sandy mud at intermediate depths, Two of the most conspicuous species are the heart urchin *Echinocardium cordatum* and the brittlestar *Amphiura filiformis* and the community has been named after these indicators. Other communities have been named in a similar way. However, there is still controversy over the best way to classify marine benthic communities.

An alternative method favoured by workers such as Jones sampling the benthos off the Isle of Man in the Irish Sea used physical characteristics of the environment, a major division being hard or soft substrates and in the latter group grain size of the sediment. This led some to suppose that there was little evidence that large assemblages of animal species interact as a unit or a 'superorganism' but are governed simply by the physical attributes of the environment. A third method of classifying communities includes the type of metabolism used. This involves looking at the amount of energy tied up in those organisms comprising the community and the flow of energy through the system. The type of community analysis chosen by researchers may depend also on the area in which the work is carried out.

If there are sharp physical boundaries then discrete communities often coincide with the physical zones. In areas of gentle gradients, such as often occur in the deep-water benthic habitat, researchers prefer the continue concept with more gradual change in the proportions of species comprising the communities over the area studied. Henery Bigelow wrote, 'When one picks up a fish one may be said, allegorically, to hold one of the knots is an endless web of netting, of which the countless other knots represent other facts, whether of marine chemistry, physics or geology, or other animals and plants. And just as one cannot make a fish-net until one has tied all the knots in their proper positions, so one cannot hope to comprehend the web until one can see its internodes in their true relationship.

Clearly, the more one knows of deep sea food webs the closer one comes to comprehending the endless web of life. The prospect is daunting, for there are many cross connections between the knots, which are continually being tied and untied. There is immense variation between one generation of knots and the next, from a few hours for a bacterium to many years for a whale. In the deep ocean our knowledge of food webs is far from adequate. Indeed, we do not even have a proper inventory of the living knots : there are still many new species to be discovered and described.

Even so, systematic and biological investigation are advanced enough to give some understanding of how the food web is made and maintained, though we may well agree that '... the single most important unresolved problem in the fields of marine biology and chemistry concerns the ... mechanism by which organic matter is transported to depth and how the energy requirements of the deep seas are met'. Our preceding surveys of oceanic plants and the animals of the pelagic and benthic environments have been partly concerned with patterns of productivity and animal means of making a living. It seems clear that benthic animals are largely dependent, whether directly or indirectly, on organic materials that have fallen from, and through, the food webs of the pelagic zones to eventually reach the bottom. First then, we must look at trophic events in the pelagic sphere.

10

Endangered Species of India

Endangered Flora of India

Introduction

The increase of human population in the last few decades demanding development in various spheres has resulted directly or indirectly in the sudden and often far-reaching disturbances in natural ecosystems. The growth of large urban areas, construction activities such as dams, buildings and roads, encroachment on vast areas of forest lands for extension of arable expanses and mining operations are examples of direct onslaughts on nature which have steadily depleted natural resources. Besides, the pollution resulting from urbanization, industrialization, excessive use of chemical fertilizers, and pesticides and also of the several giant aircrafts have in no small way contributed to eco-imbalances in soil, water and air. These imbalances have posed problems for the very survival of wild life.

Besides these, in order to satisfy his ego and establish supremacy over other life by collection hunting, trapping and poaching man has driven species on the verge of extinction. Extinction has been the destiny of a great number of plant species including several unique and irreplacable varieties. Some of these have disappeared from the earth in nature's own process of evolutionary change. But for many others, extinction has been caused by man because of his inadvertence as well as ignorance about their economic potential and ecological functions. So far only a small fraction of the diverse populations of plants is known to us and a large majority still remains unidentified. Even in respect of familiar species in many cases, little more is known than their appearance and location. Through indiscriminate exploitation, destruction of habitats, spread of harmful chemicals and introduction of alien species, a number of plants have already disappeared while others await a similar fate, at times, even without our being aware of their existence. Thus human interference has been, more often than not, responsible for depletion of plant resources and consequent decline of genetic diversity.

Then according to an estimate of the Threatened Plants Committee of the International Union for Conservation of Nature and Natural Resources (IUCN) about 10 percent (20,000 to 30,000) of the world's flowering plants are reported to be "dangerously rare", or "under threat", Norman Myers, noted conservationist, concludes that at least one species is disappearing each day in the tropical forests alone and in a few more years there may well be a species lost each hour. The seriousness of the situation can be understood from the observation of Peter H Raven that a "disappearing plant an take with it 10 to 30 dependent species, such as insects, higher animals and even other plants." Realizing the ecological importance and economic utility of plant resources, there is growing concern throughout the world about the need for conservation. As pointed out by Otto Frankle, the well known Australian geneticist, it is now time for determined efforts "to keep evolutionary options open so far as we can", and to distinct from eroding the richness of genetic reserves. During the recent past such concern has found expression in the establishment of an International Board for Plant Genetic Resources, and the UNESCO launched the Man and the Biosphere (MAB) programme.

General Background to the Problem of Threat to Plant Species

The Forest administration in India is more than 100 years old. Efforts to identify plant species as a part of wildlife and to recognise the importance for providing special protection to endangered plant species have so far been negligible. While the rapid disappearance of some beautiful animals stimulated efforts to protect fauna, this has not been the case with flora. The word 'Wildlife' had till recently been considered synonymous with animal life and consequently, conservation and naturalists have their attention only to conservation of animal species. It was only in the year 1968 at an International Conference (UNESCO 1968) that the problem of conservation of floappreciated and several recommendations were made urging the International Biological Programme (IBP), the International Union for Conservation of Nature and Natural Resources (IUCN) and various international and national organisations to initiate studies in to the problems involved, particularly the problem of protecting and preserving wild fauna and flora in their natural habitat/ ecosystems by establishing nature reserves.

Later at the 10th General Meeting of the IUCN, the Survival Service Commission reviewed the status of endangered species of plants and their habitats. The recent promulgations of the United States Endangered Species Act (1973), the UK. Wild Creatures and Wild Plants Act (1975), development of international conventions on conservations (Wetlands Convention. World Heritage Convention, Endangered Species Convention and South Pacific Convention) and the setting up of Biological Records Centre of the Nature Conservancy, UK. and threatened plants, orchids, cycads treewide awakening of the need for the conservations preservation and protection of floras of the world. According to the Volume 5 of Red Data Book on angiosperms started by the Survival Service Commission of the IUCN in the year 1970 (Melville. 1970), it is estimated that out of the total of three lakhs (0.3 million) species of plants in the world, over 20,000 were in the category of either endangered or vulnerable and threatened with extinction by the year 2000 A.D.

Rare Species of Plants, their Categories and Definitions

The following categories of rare plants have been recognised by the IUCN, mainly based on (i) present distribution, (ii) decline in number of population in course of time, (iii) abundance and quality of natural habitats, and (iv) biology and potential value of the species. These are defined at follows for purposes of conservation.

Endangerd (E)

Texa in danger of extinction and whose survival is unlikely if the causal factors continue operating. Included are taxa whose numbers have been reduced to a critical level or whose habitats have been so drastically reduced that they are deemed to be in immediate danger of extinction.

Vulnerable (V)

Taxa believed likely to move into the endangered category in the near future if the causal factors continue operating. Included are taxa, of which most or all the populations are decreasing because of over-exploitation, extensive destruction of habitat or other environmental, disturbance; taxa with populations that have been seriously depleted and whose ultimate security is not yet assured; and taxa with populations that are still abundant but are under threat from serious adverse factors throughout their range.

Rare (R)

Taxa with small world populations that are not at present endangered or vulnerable, but are at risk. These taxes are usually localised within restricted geographical areas or habitates or thinly scattered over a more extensive range.

Threatened (T)

Threatened is used in the conservation context for species which are in one of the categories Endangered, Vulnerable and Rare. Some species are marked as threatened where it is known that they are Endangered, Vulnerable or Rare, but there is not enough information to say which of these three categories is appropriate.

Out of Danger (O)

Taxa formerly included in one of the above categories, but which are now considered relatively secure because effective conservation measures have been taken or the previous threat to their survival has been removed.

Indeterminate (I)

Taxa that are suspected of belonging to one of the first three categories, but for which insufficient information is currently available.

Main Causes of Threat to Species

It will known now that several plant species have become extinct due to certain natural phenomena, such as land upheavals, volcanic eruptions, glaciation, protracted periods of rain or draught, spreading of desert lands, forest fires, and eutrophication in the geological past. While such natural processes in the past had no doubt led to the extinction of flora, the resulting new environmental conditions had also resulted in the evolution and speciation of new flora and migration of floral elements. But, in recent times man with his anthropopogenic associates and other factors or practices such as fire or 'slash and burn' for shifting cultivation (also called as 'jhum' or 'podu' cultivation in India), grazing by cattle and by several other mechanical means, has suddenly accelerated disatrous condition in natural ecosystems. Besides, commercial exploitation of entire plants, roots, rhizomes, tubers bulbs seeds and fruits has been the prime cause of depletion of more important wild economic, plants throughout the world for lucrative financial gains, in trade which flourishes body by legal and illegal means. Rauvolfia serpentina, Coptis teeta, Dioscorea spp, and Podophylum; hexandrum serve as good examples.

Area Vulnerable to Rarity and Extinction of Species in India

In India, the maximum threat of plant species lies in areas which are subjected to considerable development activity e.g., forests which lie in close proximity to villages and agricultural lands. Such areas occur in the entire length and breadth of the Himalayas, the Western and Eastern Ghats of peninsular India, the Vindhya and Satpura Hill ranges, and the densely forested hilly region in Assam, Nagaland, Meghalaya, Manipur and Mizoram in Eastern India. These regions incidentally abound in many tourist spots and pilgrimage centres, which attract large number of people who cause damage to the flora. Besides, enemies which are confined to narrow, restricted areas or specialized ecological niches, also in some respects fall in the category of threatened plants.

Some phytogeographers believe that endemic species or genera are the survivours of the one time widely distributed groups which are now in the course of gradual extinction. Other maintain that they are newly evolved and recent forms still in the process of extending their geographical range of distribution. In such cases, due to their rarity they are to be considered as threatened plants, because any disturbance or imbalance in their narrowly confined ecosystem/habitats could mean extermination of the species. Insolation, as in the case of island floras, and natural barriers such as mountain ranges are considered main factors resulting in endemism. India has a very rich element of endemics in its flora.

According to Chatterjee (1940) Indian subcontinent has about 61.5 per cent to endemic flora with about 7,000 endemic species and 134 endemic genera. Of these the Himalayas and the Khasi Hills account for about 3,000 and Deccan peninsula for about 2,000 endemic species. These figures speak eloquently of the great need for protecting the endemics. This can be dome effectively by carefully analyzing the floristic composition of the various phytogeographical units of India flora and by selecting suitable natural forests in these units and preserving them as 'Biosphere Reserves'. Sapria himalayana, Uvaria lurida, Alcimandra cathcartii, Magnolia gustavii, M. pealiana, Pachylanax pleiocarpa, Nepenthes khasiana, Dicentra royler; several species of Primula and Rhododendron and the Lady's Slipper oschids. Paphiopedilum insigne. P.hirsutissimum, P. fairienum, P. specerianum and P. venustum, are some of the noteworthy endemics of the Himalayas and Khasi Hills. Antiaris toxicaria, Companula cytinoides, Pedicularis perrotter, and the some species of the Podostemaceae are endemic to Western Ghats or the Nilgiri Hills in South India.

Himalaya and Eastern India

1. Angopteris evecta Hoffm (Angiopteridaceae) Terrestrial fern. Widely distributed in India, Sri Lanka and sepals Asia, sporadic in occurrence.
2. Aglaia perviridis Hiern (Meliaceae) Tree distributed in Khasi Hills, Meghalaya rare occurrence.
3. Abies delavay. Franchet (pinaceae) Tree A Chinese Fir recorded from Arunachal Pradesh, rare occurrence.
4. Anacolosa ilicoides Mast (Olacaecae) Tree Distributed in Khasi Hills, Meghalaya, rare occurrence.
5. Adinandra griffithu Dyer (Theaceae) Tree Distributed in Khasi Hills, Meghalaya rare occurrence.
6. Ambliyanthus multiforus Mez (Myrsinaceae) Shurb Rarely reported from Assam.
7. Acanthephippum sylthetenze Lindi (Orchidaceae) Herb one A terrestrial orchid distributed in Garampani sanctuary (Assam Kashi, Orissa, Abor Hills and Sylhet (Bangladesh) rare occurrence.
8. Anoectochilus sikkimensis King & Plant (orchidaceae) Terrestrial herb, the 'Jewel Orchid'. Distributed in Sikkim Himalaya, recently reported from Garampani sanctuary, Assam, rare occurrence.
9. Aphyllorchis montana Reichb (Orchidaceae) Herb A Terrestrial orchid, petals straw-coloured purple at back and tips. Distributed in Sikkim and Khasi Hills, rare occurrence.
10. Artemisia amygdolina Decve (Asteraceae) Herb. A botanically interesting species with leaves like those of a willow (J.D. Hooker). Known only from Pir Panjal, J&K State, extremely rare in occurrence.

11. Atropa acuminata Royle ex Lindi. (Solanaceae)–Herb. The 'Indian Belladonna', distributed in Himachal Pradesh and Kashmir; rare occurrence; exploited as medicinal plant.
12. Arundina graminifolia (D. Don) Hochr (Orchidaceae) Terrestrial, perennial plant. The Bamboo orchid with highly ornamental flowers. Distributed in Sikkim, Khasi Hills and Arunachal Pradesh, rare occurence, horticultural value, exploited in trade.
13. Astragalus strobiliferus Royle (Papilionacea)–Herb. The only member of the Tragacanth group of Himalayan Asttragali occurring in W. Himalaya: rare occurrence.
14. Aconitum deinorrhizum Stapf (Ranunculaceae) Herb One of the important Indian Aconites of medicinal value originally discovered from Bashahr Himalaya are recently reported from Bhadrawah, J&K State rare occurrence, medicinal exploited for made.
15. Arachnanthe clarker (Reicbb ff) Rolfe (Esmeralda clarker Reichb f) (Orchidaceae) Eplypytic Robust orchid with large, attractive, yellowish-brown flowers. Occurs only in Sikkim and Bhutan, rare occurrence.
16. Arachnanthe cathcartii (Lindi) Benth (Esmeralda cathcartii (Lindi) Reichb (Orehidaceae).

    Epiphytic Attractive with large brownish-red flowers. Distributed in East Nepal, Sikkim, Bhutan and Arunachal Pradesh, rare occurrence.
17. Brainae insignis (Hook) J. Sm. (Blechnaceae)–Terrestrial fern. Distributed only in Khasi Hills in India; very rare; also occurs in Malay Peninsula.
18. Botrychium virginianum Sw (Ophiolossaceae)–Perennial. A pretty fern of botanical interest. Distributed in the Himalayas (Kumaon to Arunachal Pradesh), Khasi Hills, Peninsular India and Sri Lanka, rare occurrence.
19. Balanophora dioica R. Br. (Balanophoraceae)–Herb, Clump-forming, fleshy, club-shaped parasite on tree roots in dense forsts in the entire Himalayan region; of extreme rare occurrence; botanical interest (Also other species of Balanophora).
20. Cyathea gigantea (Wall ex Hook). Holtt, (Cyatheaceae)–The giant tree fern of botanical interest. Widely distributed in

hilly regions of eastern and peninsular India, but nowhere abundant; exploited for starch by the local people; trunk used for growing epiphytic orchids.

21. Catamixis bacchariodes T. Thoms. (Asteraceae)–Herb. An enmic species restricted to rock cliffs in U.P. Siwaliks; rare occurrence; botanical interest.
22. Captis teeta wall (Ranunculaceae)–Herb. Distributed in Mishmi Hills, Arunachal Pradesh, rare occurrence; heavily exploited in trade for its medicinal rhizomes– 'Mishmi Teeta' now under cultivation.
23. Colchicum luteum. Baker (Liliaceae)–Herb, Endemic to N.W. Himalaya; rare occurrence; exploited in trade for its valuable alkaloid Colchicine.
24. Cymbidium macrohizon Lindi, (Orchidaceae)–The only terrestrial leafless India Cymbidium. Distributed in N.W. Himalaya, Sikkim, Khasi and Naga Hills; rare occurrence; horticultural importance.
25. Camellia caduca C.B.Cl.ex Brandis (Theaceae)–Shurb. Distributed in Sikkim, Bhutan, Arunachal Pradesh and Khasi Hills; rare occurrence; botanical interest.
26. Cypripedum cordigerum D. Don (Orchidaceae)–Terrestrial herb, highly ornamental "Lady's Slipper Orchid", one reported abundant at Guimarg (Coverntry's Wild flowers of Kashmir, 1923), but has been sighted only once in the course of five visits (K.S. Sahni). Also scarce in Western Himalaya and Tehri-Garhwal region; extremely rare; threatened due to trade.
27. Cypripedium elegans Reichb f. (Orchidaceae)–Terrestrial herb; flowers brownish attractive. One of the ornamental "Lady's Slipper Orchids". Distributed in lachen Valley, Sikkim and eastern Tibet rare occurrence, threatened due to trade.
28. Cypripedium himalacicum Rolfe (Orchidaceae)–Terrestrial herb; flowers showy, reddish or purplish-red. Distributed in Western Himalayan, Sikkim and Bhuttan, rare occurrence; threatened due to trade.
29. Cypripedium macranthon Sw. (Orchidaceae)–Terrestrial herb with attractive large brownish-purple flowers. Distributed from Garhwal to Sikkim; rare occurrence; threatened due to trade.
30. Dendrobium densiflorum Wall ex Lindl (Orchidaceae)–Epiphyte. A highly ornamental orchid distributed in the

tropical and subtropical forest of Nepal, Sikkim, Arunachal Pradesh and Khasi Hills but nowhere common; exploited in trade.

31. Dianthus Coschemiricus Edgew, (Caryophyllyceae)–Herb apparently endemic in Kashmir, rare occurance. (D. Chinesis L, is cultivated in India for its perfume-yielding flowers).
32. Didiciea cunninghamii King et Prain (Orchidaceae)–Terrestrial herb. An interesting monotypic orchid originally found in Sikkim and recently collected in Garhwal Himalayas after a lapse of seven decades; rare occurrence; botanical interest.
33. Discorea deltoidea wall, ex kunth (Discoreaceae) Hioscorcuceae–Herbaceous climber, Western Himalaya; becoming scarce; threatened due to over-exploitation from the wild for its medicinally important tubers.
34. Dipteris Wallichii (R.Br.) Moore (Dipteridaceae) m–Perennial herb. A fern of botanical interest. Distributed in Khasi and Jaintia Hills; recently discovered in Subansiri, Arunachal Pradesh; rare occurrence.
35. Dischidia benghalensis Colebr (Asclepiadaceae)–An interesting climber with fleshy leaves. Eastern Nepal, Sikkim, Assam rare occurrence; botanical interest.
36. Dischidia rafflesiana Wall (Asclepiadaceae)–A Climber with pitcher-like leaves, Assam; rare occurrence; botanical curio.
37. Drosera burmanns vahi (Droseraceae) Herb. The well-known insectivorous Sundew' plant. Also the other species D peltata Sm. Both distributed from Himalayas to the east and south, scarcely distributed; botanical interest.
38. Drosera indica inn (Droseraceae)–Herb. In marshy habitats. Distributed in Deccan peninsula, upto Chota Nagpur, recently reported from Birbhum (W. Bengal; sporadic); botanical interest.
39. Elaeocorpus prunifolius (C. Muell). Mast (Elaeoscarpaceae) Tree distributed in Khasi Hills and Manipur; rare occurrence; botanical interest.
40. Eremostachys superba Royle ex. Benth (Lamiaceae)–An attractive large herb on the slopes of Dehra Dun Siwaliks. Threatened due to crosion of habitat.
41. Eria crassicaulis Hook f. (Orchidaceae)–Epiphyte. An attractive orchid with long recemes and pale purple-banded

flowers. Distributed only in Kashi Hills; fare; horticultural value.

42. Galeola lindleyana Reichb. f. (Orchidaceae)–Tall, attractive saprophytic orchid with tomentose yellow flower. Sikkim, Kashi and Naga Hills. Also another species of G. faciconeri Hk. f. with deep yellow flowers, is found only in parts of Arunachal Pradesh; rare occurrence; botanical interest.
43. Gentiana kurroo Royle (Gentianaceae)–A herb with beautiful, large purple blue flowers. Western Himalaya rare occurrence, medicinally important; exploited in trade.
44. Hedysarum cachemirianum benth ex Baker (Papilionaceae)–Herb. An attractive plant Endemic in Kashmir; rare occurrence; botanical interest.
45. Helminthostachys zeylancia (Linn) (Helminthostachyaceae)–Terrestrial, perennial fern. Distributed in Submontane parts of north-east and peninsular India and deltaic Bengal, recently collected in Pilibhit district, U.P. sporadic in distribution, botanical interest.
46. Helwingia hinalaica Hook. f. and Thoms ex C.B.C.I.: (helwingiraceae)–Shrub. An attractive plant with flowers. Eastern Himalaya and Khasi Hills; rare occurrence; of great botanical interest.
47. Iiex embelioides Hook. f. (Aquifoliacease)–Small tree. Distributed in Khasi Hills; rare.
48. Iodes hookeriana Baill (Lcacinaceae)–Woody climber with attractive penicles of yellow-red drupe's Kashi Hills (and Chittagong Hills Bangladesh); rare occurrence botanical interest; horticultural value.
49. Lavatera kashmiriana Camb (Malvaceae)–Herb. 'The Kashmir Hollyhock' with pink flowers; endemic in Kashmir, scarce.
50. Lespedeza elegans Camb (Papilionaceae)–Endemic in Kashmir; sporadic in occurrence.
51. Loropetalum Chinese Olive (Hamamelidaceae)–Small tree; distributed in Assam and Arunachal Pradesh; scarce.
52. Magnolia griffithii Hook. f. & Thoms (Magnoliaceae)–Tree Endemic in Upper Assam; rare occurrence.
53. Magnolia gustavi king (Magnoliaceae)–Large tree. Endemic in upper Assam; rare occurrence.

54. Magnolia pterocarpa Roxb (Magnoliaceae)–Tall tree. Scarcely found in Assam and Arunachal Pradesh, Hutchinson (Evolution and Phylogeny of flowering plants, 1969) considers this palnt to be perhaps the most ancient species of living flowering plants. Of considerable botanical interest.
55. Meconopsis betonicifolia Franchet (Papaveraceae)–Perennial herb. Blue poppy of the Eastern Himalaya also found in S.E., Tibet, Northern Burma and Yunnan; very rare in the wild; exploited for trade; introduced into Cultivation in key Botanic Gardens in 1924 and now growth in Europe.
56. Nardostachys grandifloro DC (Valerianaceae)–Perennial herb. The Himalayan spikenard with aromatic rhizomes (Jatamansi), Alpine Himalayas rare occurrence; exploited in trade.
57. Nepenthes Khasiana Hook. f. (Nepenthceae)–Climbing undershrub. The Indian insectivorous pitcher plant. Endemic is Khasi and Garo Hills; rare; botanical curio. The populations have dwindled due to excessive collections by traders and students of botany.
58. Iax nana Wall (Olacaceae)–Tree, Distributed in Nepal, Darjeeling, recently reported from Assam; rare and sporadic in occurrence.
59. Ormosia glauca Wall (Papilionaceae)–Tree. Distributed in, Nepal, Darjeeling recently reported from Assam; rare and sporadic in occurrence.
60. Osmunda regalis Linn. (Osmundaceae) Perennial shrub. The Royal Ferm. Scattered in Kumaon Eastern Himalayas, Khasi Hills, Pachmarhi Hills, and peninsular India, but nowhere abundant; botanical interest.
61. Paphiopedilum druryi (Bedd.) Pfitz. (Orchidaceae)–Terrestrial herb, flowers yellow-green with reddish-purple blotches and streaks. Endemic in travancore Hills, W. Ghats rare occurrence.
62. Paphiopedilum fairleanem (Lindl) Pfitz. (Orchidaceae)–Terrestrial herb. The "Assam Lady's Slipper Orchid", with large attractive greenish or cremy, reddish and purple-veined flowers. Endemic in Kameng (Arunachal Pradesh and Bhutan; very rare occurrence threatened due to habitat destruction and indiscriminate collections for trade.
63. Paphiopedilum hirsutissum (Lindl) Pfitz, (Orehidaceae)–Terrestrial herb with large violet-purple flowers of ornamental

value. Endemic in Khasi Hills, becoming extremely rare due top indiscriminate collections from the wild.

64. Paphiopedilum insigne (Wall) Pfitz. (Orchidaceae)–Terrestrial herb with attractive greenish purple spottted flowers. Sparsely distributed in E. Himalayas, Khasi Hills and Nepal; becoming every scarce due to habitat destruction and exploitation in trade.
65. Paphiopedilum spicerianum (Reichb.) Pfitz. (Orchidaceae)–Terrestrial Scapigerous herb; flowers with purplish sepals and yellow or olive green, red-spotted petals and brownish crimson tinged lip. Endemic in cachar Hills, Assam; rare occurrence.
66. Paphiopedilum venustum (Lindl) Pfitz. (Orchidaceae)–Terrestrial herb with blotched; leaves; flowers sowny and ornamental with white dark-striped sepals and greenish puple wartypetals distributed in tropical Sikkim Himalayas and Meghalaya, becoming rare due to clearing of forests and over-exploitation from the wild.
67. Paphiopedilum villosum (Lindl) Pfitz. (Orchidaceae)–Terrestrial herb, flowers large, ornamental publish-brown, distributed in Mizoram (Lushal Hills) and Moulmein, Tenasserium (Burma); extremely rare and threatened due to indiscriminate collections from the wild.
68. Phyllostachys bambusoides Sieb and Zucc (Proaecae)–Shrub. A Sino-Japanese bamboo, distributed in Arunachal Pradesh; rare occurrence.
69. Picea brachytyla Pritz (Pinaecae) Tree. A Chinese spruce with restricted distribution in Arunachal Pradesh: also occurs in N. Burma, Western Hupeh and W. Yunnan (China).
70. Platycerium wallichii Hook (Polypoyllaceae)–Epiphytic fern. Recently collected from Manipur; very rare; earlier known from Malay Peninsula, Tenasserim.
71. Pleione humilis D. Don (Orchidaceae)–Herb. A pretty orchid distributed in Khasi Hills and Arunachal Pradesh, recently reported from Kumaon, W. Himalayas; rare occurrance; exploited in trade.
72. Podophyllum hexandrum Royle (Podophyllaceae)–Herb. The Himalayan many-apple widely distributed, but nowhere abundant, Heavily exploited for its medicinal value.
73. Populus gamblei Dode (Saliexceae)–Tree. 'Poplar distributed in Eastern Himalayas, rare.

74. Potameria paradoxa (Hook) Koster (Lauraceae)–Tree. A curious plant with 1-celled anthres; only once collected from Bhutan; rare occurrence; botanical interest.
75. Psilotum nudun (Linn). Beauv (Psilocaceae)–Perennial herb. A fern-ally of botanical interest. Widely distributed in India but nowhere common, botanical interest.
76. Rauvolfia serpentina (Linn). Benth ex Kurz (Apocynaceae)–Shrub. A very important medicinal plant of widespread distribution. Now scarce in the wild due to over-exploitation; cultivated, 7.
77. Renanthera imschootiana Rolfe (Orchidaceae) Epiphytic herb. A climbing orchid with showy yellow red-spotted petals and red-flushed lateral sepals. Distributed in Assam region (also in Indo-China); rare occurrence of ornamental value.
78. Rheum nobile Hook. f. and Thoms (Polygonaceae) Perrenial herb. The "rhubarb" of eastern Himalayas (400-4500 m); rare; exploited for its leaves used as substitute for tobacco.
79. Rhododendron arizelum Balf and Forr. (Ericaceae)–Tree with highly ornamental rose-pink flower bunches, Discovered as new to India for Subansiri district. Arunachal Pradesh; rare occurrence; not reported from other parts of the eastern Himalayas.
80. Rhododendron dalhousiae Hook. f. (Ericaceae)–A shrub by rhododendron with attractive large flowers; occurs in Eastern Himalayas, rare occurrence; exploited for horticultural purpose.
81. Rhododendron edgeworthii Hook f. (Ericaceae)–An Epiphytic ornamental shrub, reported from Arunachal Pradesh; rarely collected; exploited for horticulture value.
82. Rhododendron nivale Hook f. (Ericaceae)–Epiphytic herb. The smallest of rhododendrons only 5 cm. high, distributed in eastern Himalayas at an altitude of 5000 m; rarely collected: botanical interest.
83. Rhododendron natallii Booth (Ericaceae)–Tree or epiphytic shrub; flowers large white with creamy base, sweet scented. Distributed in Bhutan and Daphala Hills, Arunachal Pradesh; rarely collected.
84. Rhododendron Santapaui Sastro et. al. (Ericaceae)–Pretty epiphytic shrub with white starry flowers. Collected so far

only from Subansiri district, Arunachal Pradesh; very rare; endangered due to clearing of forests.

85. Rhododendron stenaulum Balf. W.W. Sm. (Ericaceae)–Tree. flowers white, fragrant. Discovered as new to India from Apathani valley, Subansiri District, Arunachal Pradesh. Only onc tree was seen; rare occurrence; threatened due to the clearing of forests in the Valley.
86. Rhus hookeri Sahni and Bahadur (Anacardiaceae)–Small tree. Endemic in Eastern India; rare occurrence.
87. Sapriya himalayana Griff. (Rafflesiaceae) A root parasite of botanical interest, flowers ca 15 cm across. First collected in 1836 from Mishmi Hills, then collected after a century from Aka Hills, recently collected from Lohit district; distributed only in Arunachal Pradesh; very rare; of considerable botanical interest.
88. Saussurea bracteata Decne, (Asteraceae)–Herb. A delicately fragrant species of western Himalayas; rare occurrence.
89. Saussurea gnaphalodes (Royle) Sch-Bip (Asteraceae)–A curious woolly herb of high altitude slopes of western Himalayas; rare occurrence.
90. Saussurea lappa C.B.C.I. (Asteraceae)–Herb. Endemic to Kashmir and Himachal Pradesh. Extensively exploited from the wild in the past; and needs protection in the wild. The 'Kuth' of commerce now under cultivation.
91. Schizaea digitata Sw. (Schizaeaceae)–A perennial fern with discoontinuous distribution in Kashi Hills. Sri Lanka and Malaysia; rare occurrence; of botanical interest.
92. Tetracentron siense Oliv. var himalense Hara and Kanai (Tetracentraceae)–A primitive vasselless discotyledonous plant. Distributed in eastern Nepal. Bhutan and Arunachal Pradesh; rare occurrence; of considerable botanical interest.
93. Thylacospermum rupifragrun Schrenk (Caryophyllaceae) Herb. Montypic cushion-forming plant of alpine Himalaya: rare; of botanical interest.
94. Vanda coerulea Griff (Orchidaceae)–Epiphytic. 'Blue vanda' of Asia, most beautiful of vandas. Distributed in Khasi and Jaintia Hills and Naga Hills: recently discovered in Tirap district, Arunachal Pradesh. Excessively exploited from the wild for ornamental purpose; now very scarces in the wild.

95. Vanda pumila Hook. f. (Orchidaceae)--Epiphytic herb. Occasionally found in Sikkim and Bhutan; rare in the W. Himalaya.
96. Vanilla pilifera Holtt. (Orchidaceae)–Climbing orchid with fleshy leaves and creamy flowers. Garmapani Sanctuary (Assam); very rare; also distributed in Malaya.
97. Viola Falconeri Hook. f. and Thoms (Violaceae)–A pretty herb. Endemic in Kashmir; rare occurrence.
98. Zanthoxylum scandens BI. (Rutaceae)–Straggling shrub, distributed in Khasi Hills; rare occurrence.

Rajasthan and Gujarat

99. Commiphora wightii (Arn.) Bhandari (Burseraceae)–Small tree. Exploited in the past for its gumresin from the ark. Distributed in Rajasthan and Gujarat in India; reportedly becoming scarce due to over exploitation; presently cultivated.
100. Helichrysum cutchicum (C.B.Cl) Rao et Deshpande (Asteraceae)–herb. Endemic in Gujarat; rare occurrence.
101. Hyphaene dichotoma (Baker) Furtado (H. Indica Becc). (Arecaceae)–The Indian doum palm, curiously branching, Distributed only in W. coast regions; of botanical interest.
102. Rosa involucrata Roxb. (R Lyellii Lindl. 3 (Rosaceae)–Subscandent shrub. Widely distributed in India: reportedly becoming scarce at Mt. Abu in Rajasthan due to clearing of forcoto.

Andaman and Nicobar Islands

103. Ailanthus kurzii Prain (Simaroubaceae)–Tall tree. Endemic in Andaman Islands; becoming rare due to clearing of forests and not collected daring the recent botanical explorations in these islands.
104. Canarium manii King (Burseraceae)–Tree. Distributed in Middle Andaman Island; Parkinson reports is as not uncommon but his species has not collected during the recent botanical explorations; probably becoming rare.
105. Dipterocarpus kerrii Kink (Dipterocarpaceae) Tall tree, Occurs in South Andaman. "Seen only in the Goplakabang Valley, where a sample plot of this has been made" (Parkinson, For Fl. Andaman Island, P. 93, 1923). This species has not been collected during the recent botanical explorations in Andaman Islands; rare.

106. Hippocratea nicobarica kurz (Hippocateaceae)–Climbing shrub, distributed in tropical forests of eastern coast of Katchall Island (Nicobar group); rare occurrence; not collected during the recent botanical explorations.
107. Lagerstroemia hypoleuca Kurz (Lythraceae)–Tree. Widely distributed in Andamans. This species is one of the important timber yielding trees and has been exploited much; becoming scarce in wild.
108. Myristica andamanica, Hook. f. (Nyristicaceae)–Tree. Distributed in middle Andaman, South Andaman and Porlob Islands; becoming scarce due to clearing of forests. Not collected in the recent botanical explorations.
109. Ophioglossum pendulum Linn. (Ophioglossaceae)–Epiphytic herb with long ribbon like fronds. Recently reported from Campbell Bay, Great Nicobar Island; rare occurrence of botanical interest.
110. Podocarpus nerifolius D. Don (Podocarpaceae)–Tree. The only confier of the Andamans, although widely distributed; exploited for its valuable timber, becoming scarce in these Islands.
111. Psilotum complanatum Sw. (Psilotaceae)–Epiphytic pendulous herb, recently reported from Great Nicobar Island; rare occurrence; of botanical interest.
112. Symplocos odoratissima (B.) Choisy ex. Zoll (S. Chengapae Raizada and Sahni) (Symlocaecae)–shrub or tree. Distributed in Great Nicobar Island; rare occurrence. Also occurs Indonesia, Malaya, Singapore and Philippines.
113. Uvaria nicobarica Raizada and Sahni (Annonaceae)–Woody climber, Endemic in Great Nicobar Island; rare occurrence.

Peninsular India

114. Anemia tomentosa Sw. (Schizaeceae)–Terrestrial fern. Distributed in Nilgiri. Annamalai, Palney Hills; rare occurrence.
115. Apama barberi Gamble (Aristolochiaceae)–Shrub, Endemic in Tirauelveli Hills; rare occurrence.
116. Bentinkia condapanna Borty ex. Roxb (Arecaceae)–A reed-like palm. Endemic in Kerala and Tirunelveli Hills; of botanical interest; rare occurrence.
117. Ceropegia fantastica Sedgw (Asclepiadaceae) Slender climber, described by Sedgwick in 1921, based on a single collection

from N Kanark, and known only form Holotype (Kew) until 1963, when this species as again collected in Goa (BSIWC); very rare.

118. Cycas beddomei Dyer (Cycadaceae)–Palm like tree. Endemic to Cuddapah Hills, S. India; rare.
119. Dioscorea wightii Hook. f. (Dioscoreaceae)–Twinning climber with tuberous roots. Endemic in Tirunelveli Hills and recently collected after 100 years from type locality; very rare.
120. Entada pursaetha DC. (Mimosaceae) Large woody climber with gigantic pods of botanical interest. Distributed in Himalayas. Assam and W. Peninsula; becoming rate due to large scale clearing of forests.
121. Frerea indica Dalz. (Asclepidaceae) undershrub. Endemic in W. Ghats. S. India, rare occurrence.
122. Gnetum ula Brongn. (Gnetaceae)–Extensive woody climbing gymnosperm of botanical interest. Chiefly distributed in W. Ghats: becoming rare due to large scale clearing of forests.
123. Hora-wightii Hook. f. (Asclepidaceae)–Large perennial climber, often becoming totally epiphytic Endemic to W. Peninsula becoming scarce due to clearing of forests and habitat destruction.
124. Lilium neilgherrense Wt. (Liliaceae)–The Nilgiri-lily. Endemic to the Niligiri and a few high mountain ranges is S. India; rare occurrence.
125. Lobelia nicotianaefolia Roth ex Roem. and Schult. (L. excelsa Leschen) (Lobeliaceae)–Herb, distributed in W. Ghats; rare occurrence.
126. Loeseneriella bourdillonii (Gamble) Rammam. (Hippocratea bourdillonii Gamble) (Hippocrrteacea)–Climbing shrub. Endemic to S.W. India; rare occurrence.
127. Manisuris devergens (Hack) Kuntz. (Poaceae) Tufted grass. Endemic to peninsular India: collected aftered 130 years from Coorg district; rare occurrence. All the 12 species of Manisuris found in India and are endemic; ten of them confined to South India.
128. Piper barberi Gamble (Piperaceae)–Climbing shrub with pinnately-veined leaves; endemic in Tirunelveli Hills; botanical interest; rare occurrence.

129. Podocarpus wallichianus (Podocarpaehae)–Tree. The only native conifer of peninsular India, also occurs in E. Himalayas, Khasi Hills and Nicobar Islands; rare occurrence.
130. Pterocarpus santalinus Linn. f. (Papilionaceae)–Tree. Endemic in Cuddapah district, S. India; the Red Sanders of Commerce, exploited much in the past for its beautiful red wavygrained wood. Export already banned.
131. Pterospermum obtusifolium Wt. (Sterculiaceae)–Tree. Distributed in Malabar and Karnataka; rare occurrence.
132. Satalum album Linn. (Santalaecae)– Tree. Usually Semi-parasitic. Occurs in peninsular India. Sandalwood of commerce, much exploited in trade, becoming rare (now cultivated).

Gangetic Plain

133. Aldrovanda vesiculosa Linn. (Droseraceae)–Aquatic herb. probably monotypic, carnivorous plant of botanical interest. Reported from salt pans near Calcutta; recently reported from Manipur and Bangladesh. Its habitat near Calcutta (Salt Lake area) is threatened due to reclamation.

Gaps in our Knowledge Rare and Endangered Indian Plants

The main lacunae in our knowldege with regard to this subject lies in the present actual status of the species i.e., the exact number and location of populations of these species occurring in the wild, quantities of the numbers collected for economic exploitation or other purposes; their capacity to reproduce and propagate themselves in nature and their population's or representatives in cultivation under artificial conditions in farms or experimental gardens. Such observations can be made by repeated visits to the field and need concerned efforts of several botanists and constant monitoring. This naturally an enormous task for a country like India of the size of a sub-continent.

A recent close study of the orchid genus coelogyne serves as a good example four-species of the genus viz. O. albo-lutea, C. treutleri, C. brevifolia and C. assamica, are known only by their mention in literate and types (specimens/illustrations). These could not be located even in their type localities, inspite of several recent botanical explorations and are believed to have become extinct. Excepting the types, even their preserved specimen are not available in any herbarium. Only such intensive library, field and herbarium studies can reveal the real status of our threatened taxa. Similarly the recent discovery of the Indian pitcher plant, Nepenthes khasiana in Garo Hills near Tura a locality other than its own type locality near Jowai, also highlights the need for intensive botanical explorations and constant monitoring of the floras in the country.

Action Taken to save the Rarities of Indian Flora

On the recommendations of the Botanical Survey of India and other concerned agencies, the Government of India has imposed restrictions or a complete ban on the export of certain species. The Wildlife Wardens/Chief Conservations of forests in such State have been appointed as custodians of flora and fauna to check illegal trade practices in these species. They can be exported only when certified by the specified authority (presently the Director, Forestry & Wildlife). India is one of the signatories to the Convention on International Trade in Endangered Species of Wild Fauna and Flora (CITES, 1973), it has been ratified so for by 67 countries of the world, for strict regulation and control to trade. According to the Convention, if a species is even smuggled out of a country.

The country destination will co-operate by obstructing the import of the consignment. The National committee on Environmental Planning and Coordination .(NCEPC) and the National Committee on Man and Biosphere (MAB) have also been concerned with the protection habitats which undisturbed natural vegetations several such areas have been identified for conservation and preservation as 'Biosphere Reserves', throughout the country. It is also proposed to set Germ Plasm Banks and Gene Sanctuaries to protect the areas rich in wild relatives of certain economic and cultivated plants. The Botanical Survey of India with its main Botanic Garden at Sibpur and the regional experimental gardens established at different altitudinal situations to suit the acclimatizations of plants occuring in different ecological conditions in the country, have in their holdings rare, endemic and threatened plants.

Several species of orchids are also under cultivation and protections in the National Orchidaria at Shillong, Yercaud and Calcutta. Of these, Vanda Coerulea (Blue Vanda), Dendrobium nobile, Cymbidium grandiflorum, Renanthera and species of paphiopedilum and pleione growing in shillong deserve special mention. The Government of Meghalaya State has taken action on the recommendations of the Botanical Survey of India, and declared the Nepenthes khasiana habitats as 'protected areas' to preserve the plants in situ.

Future Programme of Work in the Botanical Survey of India

In order to fill the gaps in our knowledge on threatend plants the following studies are being undertaken by the Survey.

1. Collection of live plants for introduction into the botanical gardens for protection and further studies on propagation.
2. Preparation of distribution maps of rare and endangered plants.
3. Preparation of an identification manual of plants banned for export of facilitate proper identification and for education public.
4. Fresh explorations to areas for collection of plants and data on rare/endangerd species with regard to the present distribution frequency and phenological data.
5. Preparation of an illustrated took on rare and endagered plants.
6. A review of literature and study of herbarium specimens of endangered or rare species deposited in the various herbana of the country to collect data on the distribution, abundance and frequency.

Endangered Fauna of India

As many as 500 million kinds of plant, animal and microorganism have made this planet home since life began over 3.5 billion year ago. Today, there are only 5 million to 10 million species alive we do not know exactly how many, because there are many biologically uncharted areas such as the tropical rain forests where some estimate that such as the tropical rain forests, where some estimate that over 90% of the living organisms remain unclassified. Thus, since life began, about 490 million species have become extinct. As study of the history of life from an evolutionary prespective tercher us that biological extinction is a fact of life. As species evolve, the old give way to the new.

Many of the so called extinct species are represented today by their descendants. Thus, many species did not off as such but were simply altered during evolution. Other species vanished completely as a result of drastic climatic changes or increasing environmental resistance created by excess predation or disease. The dinosaurs, for example, perished rather abruptly all over the world 65 million years ago, possibly as a result of global cooling. Much more recently, at the end of the pleistocene epoch 10,000 years ago, many large mammals such as the sabertooed tiger and giant tree sloth vanished from North America. Their extinction resulted either from climatic changes at the end of the Ice Age of possibly from Siberia. As these people moved south and east, they hunted the large mammals such as they wooly mammoth wooly rhinoceros, giant deer, ground sloth, mastodons, giant beavers, sabertoothed tigers, bioson, and elk. They undoubtedly also competed for the same food resources.

From the end of Pleistocene until now, the earth's climate has been fairly constant and relatively few plants and animals have become extinct in the 1800s, 84 animals species and subspecies are known to have vanished. By 1984 the number had risen to 184, and if trends continue, many scientists believe, 170 more species will disappear by 2000, bringing to total to more than 350 if conditions turn considerably wore. More than 700 additional species might be destroyed before the end of the century. One vertebrate species is becoming extinct every nine months, compared with a natural rate of one species every thousand years. When plants, insects, and microbes are added, the rate may be as high as one species per day.

Ecologists argue that plant extinctions will have much more profound impacts on the ecosystem the extinction of animals, because plants from the base of the food web. Some authors assert that ifthe world's population continues to grow and nations continue to destroy wildlife habitat by expanding agriculture and other activities at the current pace, the extinction rate could reach 20,000 species a year. At the rate 400,000 species of plants and animal could be wiped out in 20 years. World wildlife expert Norman Myers argues that we could lose 1 million species by 2000. Although such high rates of extinction are probably exaggerated extinction will undoubtedly increase unless we make conscious efforts to curb population growth and properly manage our renewable and nonrenewable resources.

Why Save Endangered Species?

Wildlife expert Norman Myers once wrote. "We can marvel at the colours of a butterfly the grace of giraffe, the power of an elephant, the delicate structure of a dia-tom. Every time a species goes extinct, we are irreversible improverished." Aestpeties in itself is a good argument for saving species. As we preserve our collection for future generations, so we should preserve this living art form that painters and sculptors only dream of re-creating. A second reason for protecting species is that other organisms have a right to live. Can we not guarantee them the right to exit, something that we ourselves doggedly strive for? The third reason for preserving other organisms is a matter of ethical responsibility. As a species capable of destroying the world, some, ecologists and philosophers feel, we have a moral obligation to preserve organisms with lesser powers. The fourth reason for protection is that it makes good economic sense to do so.

As Myers wrote; "From morning coffee to evening nightcap we benefit in our daily life style from the fellow species that share our One Earth home, Without knowing it, we utilize hundreds of products each day that owe their origin to wild animals and plants. Indeed our welfare is intimately tied up with the welfare of wildlife. Will may conservationists proclaim that by saving the lies of wild species, we may be saving own. From the biosphere we reap countless benefits: fish from the sea, medicines and other products from plant important plant and animal genes needed to improve domestic crops and live-stock, something for hunters to shoot at something for anglers to hook and reel in some thing for bird-watchers to peer at and research animals that help provide valuable insights into human physiology and behaviour. The ecomomic benefits of these activities can be enormous. In 1983 alone trout and salmon anglers spent an estimated $5 billion on equipment, transportation food and lodging and hunters and bird watchers spent billions on tennis shoes, supplies, books gasoline food and lodging. Billions of dollars pour into companies that cut timber or extract valuable medicines from plants.

Each time we take a prescription drug, the chance are one in two that its origin was wild plant. The commerical value of such drugs is around $15 billion per year in the United States and about $40 billion worldwide. The U.S. Department of Agriculture estimates that each year genes bred into commerical crops yieldover $1 billion worth of food. Similar grains can be documented for Canada, Great Britain, the Soviet Union, Australian, and other agricutural nations. About half of the increased productivity in corn over the past 50 years has resulted from "genetic transfusions" from wild relatives of corn or primitive cultivaters in use only in isolated regions. Some biologists argue that the current high-yielding corn, which produces 250 bushels per hectare (650 bushels per acre), couldbe increased fivefold with further genetic transfusions from its ancestors. New developments are on the horizon that may offer financial grains and a more healthful life for us. For Example, the adhesive that barnacles use to adhere to the bottoms of ships may provide us with a substance to cement tooth fillings in place.

A chemical derived from the skeletons of shrimps, crabs and lobsters (chitosanase) may serve as a preventive medicine against fungi infections. A new drug that kills viruses may be developed from a Caribbean sponge till now physicians have been virtually helpless in the battle against viruses. The future of the biological world is one of grave economic concern. The loss of fish form acid rain and over fishing results in a loss of income to owners of fishing lodges and sporting goods stores. Ecological impoverishment will keep more of us to home, diminishing income to the recreation and tourism industry. Plants add animals lost before they are explored for possible benefits will diminish our opportunities to flight disease and increase productivity.

A dozen genetic reservoirs (so-called centers of diversity) located in the tropics and subtropics are the source of virtually all commerically value plants and animal. And they provide genetic material needed in the continual battle to imporve plant and animals resistance. Loss of these centers would have a global impact on food supplies. The final reason for preserving species is to protect ecosystem stability. All living organisms are an integral part of the biosphere and provide invaluable services. These include the control of pests, the recycling of nutrients, the replenishment atmospheric oxygen, the maintenance of local climate, the control of groundwater levels, the regulation of atmospheric carbon dioxide and global climate and the control of flooding.

It is no exaggreation to say that civilization depends on the survival of the biological world. Perhaps in the short term a species lost here and there may be of little consequence for overall ecosystem stability but in the long term the cumulative effect of such losses may someday threaten our existence. Critics of the animal protection movement argue that two much effort is directed to the savining to endangered species. Why was there so much fuss over the tinysnail darter a fish imperiled by the Tellico Dam built by the Tennesse Valley Authority (TVA). Surely this one species can have no great significance to us although no irreparable ecosystem damage would result from its extinction, the question that arises is where to draw the line. Wildlife advocates argue that if we take the attitude that each species by itself is dipensable, bit by bit we will destroy the rich biological world we live in. Somewhere the line has to be drawn each endangered species in worth saving because it stops the momentum toward widespread destruction. In growth-oriented societies this momemtum may be difficult to stop, much less slow down. Therefore each hurdle put in its way becomes an important force in saving the living creatures that make up our web of life.

The Indian Scene

At one time our wild life was as remarkable for its abundance as it is even now for its variety, with tigers haunting the forests in hundreds, and antelopes the open plains in thousands upon thousands. Over the centuries it has played an important role in our economy and culture-as a source of protein for the masses, as items of trade, and as an inspiration for a rich religious lore. The wild life has gradually declined over the millenia, and very rapidly since the British took over the country. At last we have reached a critical stage at which most, if not all of our wild life appears on the verge of being totally wiped out. Only bold and imaginative action in conservation and management will enable us to save this rich heritage from an irreversible loss at this critical juncture. The Indian subcontinent may be divided into the following eight regions for our purpose:

1. The forested hill-slopes of Himalayas
2. Ladakh and adjacent regions
3. The alluvial plains of Ganges and lower Brahmaputra
4. Assam and adjacent hill tracts.
5. Punjab and the adjoining semi-arid and desert regions.
6. The Western Ghats of Sahoadris
7. The eastern peninsular tract
8. The western peninsular tract excluding the Western Ghats.

Each of these ecological regions has been subjected to distinctive demagogic pressures, commerical exploitation agricultural expansion and other modern development activities, which are so much a part of the present day industrial and agricultural economy.

Today, the position of wild animals is, in general, alarming, and many species are greatly endangered and are fighting a losing battle for survival. Happily, a general awareness has developed among the informed public about the necessity to conserve the natural ecosystems of the country and the living wild communities of animals and plants these contain.

To ensure this, certain legal steps to regulate exploitation of wildlife have been taken by the Central and state Governments. Endangered and threatened species of animals are totally portected, and barring vermin of very common species, exploitation of the rest is regulated by law under the provisions of the Wildlife (Protection) Act, 1972. To grain a full appreciation of the endangered fauna of the country, a brief account of the occurrence of different species of Mammals, Birds, Reptiles, Amphibians, and Insects in the country and their endangered forms, is presented below.

Mammals

About 350 species constitute the mammalian fauna of India of which 81 are greatly endangered.

Primates

The order primates are represented by 19 species, of which small and slender-bodies, and is restricted to the tropical rain forests and open wood land forest of South India, while the slow loris (*Nycticebus coucang*) is bigger and inhabits the denser forests of north-east India. They are persecuted chiefly for their fish, and for the supposed medicinal value of various parts of the body. Among the macaques the liontailed (*Macaca silenus*) is one of the world's most endangered primates, surviving in evergreen forests of the Western Ghats of South India, its total population is estimated to be about 800. Only the pigtailed (*M. assamensis*) occur in the dense evergreen forests of the north-eastern India, while the carbeating (M. fascicularis is restricted to a few islands) in the Nicobar group. All of them are exploited for economic purposes and the first two species for their flesh and for alleged medicinal value.

In the langur group, the Nilgiri langur (Presbvtis johni) is a multihabitat species occurring in addition to the Sholas, in the temperature evergreen forests above c 1700 mt, altitude in the Western Ghats. Its status is not as precarious at that of the liontailed macaque, it is hunted for its flesh or supported medicinal properties, the skin used for making drums, its, fur is also prized. The capped

Fig. 10.1. Hylobates (Gibbon).

(P. pileatus), the golden (P. geei) and Phayre's leafmonkey (P. phayrei), inhabit tropical and deciduous dense forests in the north eastern India. The only ape found in India is the Hoolock gibbon (Hylobates hoolock), which thrives in the hilly forests of north-eastern India. It is hunted by local tribes for its flesh and trapped for sale.

Bear

The Himalayan brown Bear (Ursus aroctos) occurs in the upper reaches of the Himalaya from Kashmir east to Sikkim, occasionally coming down to lower reaches in severe witner the Malayan sun Bear (Helarctors malayanus) inhabits the hilly terrains of north-eastern India, south of the Brahmaputra, and the sloth Bear (*Melursus ursinus*) in the well-wooded forests all over India. They are hunted

Fig. 10.2. Selenarctos (Black Himalayan beer).

for their fur and gall bladder which is supposed to have medicinal properties, also for flesh, fat and other derivatives. The family Mustelidae has 14 species in India, of which the Ermine (*Mustelidae erminea*) occurs in the western Himalaya and is hunted for its pelt; whilt the ratel (*Mellivora capensis*) is found all over the country and is killed mainly to protect poultry from it and to some extent for its skin.

The family Viverridae is represented by 15 species of which the Malabar Civet (Viverra megaspilla) occurs in the coastal district of Kerala and is feared to have been nearly exterminated; the spotted linsang or tiger-civet (Prionodon pardicolor) ranges from Nepal east of Arunachal Pradesh extending further eastword in the mountain and hill forests, but is nowhere common.

Hyaena

The striped hyaena (Hyaena hyaena) is common in the open country with low hills and ravines throughout India. It is persecuted for its habit of predating on livestock and for occasional childlifting.

Felidae

There are 15 Indian species in the family Felidae or which the tiger (*Panthera tigris*) is multihabitat species found practically throughout the country except the desert region, even upto c. 3050 mt. height in the Himalaya, unrestricted killing for trophies, skins for commerical purposes, flesh and fat for medicinal use, derivatives of body parts for their supposed magical properties, and sometimes through eradication of proclaimed animals as man-eater or Cattle-lifter, had declined its population to 1827. Through 'Project Tiger', its population has now increased to 2278 within a few years. The Indian lion (P. leo) is restricted to the Gir forest in Gujarat. The leopard (P. pardus) and the show leopard (P. uncia) are two other big cats.

Fig. 10.3. Panthera tigris (Tiger).

The former is widely distributed throughout the country in wooded forests, while the latter is restricted to higher reaches all along the Himalaya, from 2000 to 4000 mt. altitude. These have been largely hunted for commerical purposes or for the protection of livestock. Among the lesser cats, the desert cat (*Felis silverstris*), the lynx (*F. lynx*), and the caracal (*F. caracal*) are found in the arid

Fig. 10.4. Panthera paradus (Panther).

and semi-arid tracts of the western and north-western India while the jungle cat (*F. chaus*) and the leopard cat (*F. bengalensis*) are found throughout the country in small numbers. Many species of cats are hunted for their pelt. The cheetah or hunting leopard (*Acinonyx jubatus*) which once occurred in the plains and the hilly regions of northern, central and peninsular India, south to Karnataka, is already extinct from India.

Dugong

The dugong (Dugong dugong) among the Sirenia, occurs in the Gulf of Kutch, on the coast of Malbar, Gulf of Manaar and the Andaman and Nicobar Islands. Once an important dugong 'fishery' existed in the Gulf of Mannar for its flesh, but the species are now scarce there.

Rhinoceros

Among the Perissodactyla, the Great Indian one-horned rhinoceros (*Rhinoceros unicornis*) as once extensively distributed in the Gangetic plains, but is now restricted in India of North Bengal and the plains of Assam. It was on the verge of extinction in the beginning of this century when 1904, only about a dozen individuals remained in Kaziranga (Assam) and fewer in North Bengal. It was in great demand for the alleged aphrodisiac properties of its horn and almost every part of the animal including flesh was in great demand. Though the most sought after part of the rhino was its horn, its urine was also credited with antispetic properties. It became the Victim of unrestricted killing and trapping. Measures for conservation adopted since 1904 have helped to rehabilitate it, and today its population is estimated to be around 1000 in the Indian territory.

The Indian wild ass (*Asinus fiemionous*) is restricted to the saline flats of the Rann of Kutch. Its population has dwindled greatly from 5000 in 1946 to 870 in 1972. The present population is estimated to be 720. The major reasons for its decline are probably Surra disease and regular disturbance in their habitat by the saltworkers. In the family Cervidae, out of the nine species; five are endangered. These are Kashmir stag or hangul (*Cervus elaphus hanglu*) which is restricted to the northern side of the Kashmir Valley and some adjacent valleys, its population is estimated to be about 100 m Dachigam near Srinagar and another 200 elsewhere is Dessu in the Bringi Valley and Chumki, over a and Khiram in the Lidder Valley. The Swamp deep or barasingha (*C. duvauceili*) is represented by two sub-species, the swamp-dwelling (*C.d. duvaucelli*) which lives in, marshy tracts of the Tarai and Duras from northern parts of the parts of the upper Gangetic plains east to Assam. The other, the hard ground living (*C.d. branderi*) thrives on the open grassy land of Madhya Pradesh. The population of both the sub-species together in critically low now mostly confined to the Kanha National Park in Madhya Pradesh. The Alpine must deer (*Moschus sifanicus*) occur in the higher reaches of the Himalaya from Chamba Valley in the west to Sikkim in the east and the forest muck deer (*M. chrysogaster*) is found from Garhwal in the west to Arunachal Pradesh and Assam in the east.

The family Bovidae contains 21 species, of which the under mentioned 14 are endangered. The blackbuck or Indian antelope (*Antelope cervicapra*) occupies open plains, while the chinkara or Indian gazella (*Gazella dorcas bennetii*) is basically an animal of the desert, but it also inhabits the grassy plains and scrub zones of the country. The blackbuck population is about 10,000 to 15,000 in the western and southern India, while the chinkara is over 500 in Gujarat alone. The chiru of tibetan antelope (*Pantholaps hodgni*) d the Tibetan gazelle (*Procapra picticaudata*) are both found in Ladakh while the latter extends to Kumaon and Sikkim. The gaur or Indian bison (*Bogaurus*) is found in the hill forests of India. The wild yak (*Bos grunniens*) is known within the Indian limits only from Cheng chenmo Valley in northern Ladakh.

The wild buffalo (*Bubalus bubalis*) is now restricted to certain swampy areas of Assam and Madhya Pradesh, its total population is around 1000 being 500 in Kaziranga, 300 to 400 in Manas (both in Assam) and about 100 in Baster (Madhya Pradesh). The urial or shapu (*Ovis orientalis*), the Markhor (*Capra falconeri*) and the Himalayan biex (*C. ibex*) frequent the grassy patches in the high altitude regions of Kashmir, Himachal Pradesh and Punjab, and the nayan of Great Tibetan sheep (*Ovis ammon hodgsoni*) in the plateau regions from Ladakh east to Sikkim on the higher reaches of the Himalaya. The serow (*Capricornis sumatraensis*) occurs in the thickly wooded areas of the Himalaya between c. 1850 to 3050 mt. attitude from Kashmir east to Arunachal Pradesh.

Lagomorpha

The Lagomorpha is represented by four species, of which the hispid hare or Assam rabbit (*Caprologus hispidus*) is a rare and elusive species found in the grassland of the Tarai and the Duars and also in the Brahmaputtra Valley. The order Rodentia forms the largest group represented by 95 species within the Indian territorial limits, of which 11 species of flying squirrels and two of marmots are endangered, being widely exploited for their pelt as well as for their flesh. The common giant flying squirrel (*Petaurista petaurista*) is widely distributed throughout the wooded parts of the country but occurs in small numbers. The woolly flying squirrel (*Eupetaurus cinereus*) is localised in the high altitude regions of Kashmir and Sikkim. The smaller Kashmir flying squirrel (*Hylopetes fimbriatus*) is confined to Kashmir. The Himalayan marmot (marmota boback) occurs in the high altitudes of the Himalayas from Kashmir east to Sikkim, while the Long-tailed marmot (M. caudata) is restricted to the high altitudes of Kashmir within the Indian limit.

Cetacea

As many as 25 species of Cetacea are known from the Indian territorial limits and are all endangered. The Gangetic dolphin (Platanista gagetica) is found in the river system of the Ganga and Brahmaputra. Other cetaceans are marine forms. The baleen whales (Balaenoptera spp; Megapter sp.) and other whales and dolphin species (Kogia sp; Stenella spp; Sotalia spp; Steno sp: Tursiops sp; Lagenorhynehus sp; Orchaella sp, Pseudorca sp; Globicephala sp; Physeter sp: Ziphius sp; Neomeris sp; and Delphinus spp.) have been indiscriminately exploited for baleen for commerce, and for flesh, fat, bone and ambergris.

Fig. 10.5. Phocaena (Porpoise).

Birds

There are about 2100 species and subspecies of birds; known in India, of which a large number of birds are greatly endangered. Below is a brief account of these species.

Ducks and Swans

Of the ducks, geese and swans (family Anatidae), the pinkheued duck (*Rhodonessa caryophyllacea*) occurred in forest ponds and swamps along the Himalayan foothills and Terai, from Nepal east to Manipur and also appeared sporadically in places as far away as Punjab in the north and Tamil Nadu in the south. At present, it is probably extinct from India. The whitewinged wood duck (*Cairina scutalata*) breeds on trees in dense swampy forests fo Assam; Arunachal pradesh and Manipur. Migratory swans, mely, the whitling swan (*Cygnus columbianus*), the whooper swan (*C. cygmus*) and the mute swan (*C. olor*) are exceedingly rare vagrants or stragglers. They are disappearing for want appropriate habitat.

Hawks and Eagles

Most of the diurnal birds of prey belonging to the families Accipitridae, Pandionidae and Falconidae live a very precarious existence due to habitat destruction, pollution hazards, poison contamination, trapping for their tendency to prey on poultry ect. Various species of hawks (*Accipiter spp.*) eagles (*Aquilas* spp.) and falcons (*Falco* spp.) are widely distributed but in small numbers, and they are trapped for trapped for falconry. The whitebellied sea eagle (*Heliacetus leucogaster*) of the Indian coastal areas and the Bay Islands are of small population size. The *Nieobar megapode* (Megapodius freycinet), family Megapodidae is confined to the Nicobar group of islands within the Indian limits, its previous abundance has been affected by habitat destruction, increased human interference and exploitation, specially for its flesh and eggs.

Game Birds

Phasianid birds (family Phasianidae) are in great danger chiefly from human exploitation for their flesh they have always been branched as 'game birds'. Many of these are gorgeously coloured (Ornamental) and are in heavy demand from zoos and menageries for which they are trapped.

Pheasant

Bamboo partridge (*Bambusicola futchi*) restricted to open scrub and bamboo jungles from the plains to about 2000 m altitude in north-eastern India is fast loosing its habitat due to human interference. The mountain quail (*ophrysia superciliosa*) known only from a few specimens taken in the late 19th century in Kumaon is now thought to be extinct. The blood pheasant (*Ithaginis cruentus*) occurring in the Himalayas between 2000-4500 m altitude from Nepal east to Arunachal Pradesh, the western Tragopan (*Tragopan melancephala*) of the western Himalayas from Kashmir to Garhwal between 1000 and 3000 m altitude of the Himalaya from eastern Garhwal east to Arunachal Pradesh between 2500 and 4000 m altitude. The Indian peafowl (Pavo cristatus), although not threatened, needs to be mentioned since it is the National Bird of India.

Cranes

Of the cranes (family Gruidae), the eastern common crane (Grus grus Lilfordi) is a winter migrant in the larger wetlands of northern India, the blacknecked crane (Grus nigricollis), a greatly threatened bird, breeds in Ladakh at the altitude above 3000 m and presumably also in the eastern Himalayas, the Hooded Crane (Grus monacha) is a rare winter migrant to eastern Assam, the Siberion crane or the great white crane (Grus leucogeranus) is a regular winter visitor to the protected wetlands of northern Indian plains but a decidedly diminishing species. The masked finfoot (*Heliopais parsonata*, family Heliornithidae) a unique relic species, known from eastern Assam and Arunachal Pradesh in dense lowland jungle swamps add ponds. It is so rate that no authentic record of its occurrence is available for the last fifty years or so.

Bustards

Of the bustards and floricans (*family Otididea*) the little bustard (*Otis tetrax*) and the houbra or Macqueen's bustard (*Chlamydotis undulata*) are winter migrants in northern and north-western India: the great Indian bustard (*Ardeotis nigriceps*), the Bengal florican (*Eupodotis bengalenis*) Sypheotides indica) are resident species. Their previous abundance as well as range have shrunk greatly and the birds are facing extinction. Of the pigeons and doves (family *Columbidae*), the Nicobar pigeon (*Caloenus nicobarica*), restricted to the Nicobar group of islands, is fast loosing its breeding ground and is confronted with laughter by the local people.

Hornbills

The hornbills (family Bucerotidae) are large birds, and most species are threatened, chiefly due to deforestation resulting in a dearth of suitable nesting trees, and persecution by man for alleged medicinal values of their flesh, fat and bone, and for their casque and other derivatives used as ornaments, charm of in witchcraft by the tribal people. The most threatened of them are: the whitethroated brown hornbill (*Ptilolaemus tickelli*) found in the hills of eastern India in evergreen and cutover scrub forests from 300 to 1500 m altitude; the rfusnecked hornbill (Aceros ripalensis) found in central and eastern Himalayas in dense deciduous and evergreen forests of Sikkim, Arunachal Pradesh, Manipur and Mizoram; the wreathed hornbill (*Rhyticeros undulatus*) found in the tropical and subtropical evergreen forests along the Himalayan foothills in the Duars and also in the region south of Brahmaputra from the plains to on altitude of 2500 m; the Narcondam hornbill, (*Rhyticeros* (undulatus) *narcondomi*) and endemic species in the Narcodam Island of the Andaman group is highly vulnerable on this volcanic Island of the Andaman group is highly vulnerable on this volcanic Island; the great pied hornbill (*Buceros bieornis*) of the tropical and subtropical evergreen, and most deciduous forests from the plains upto 2000 m altitude, from Kumaon east to north-eastern states, and in the Western Ghats from Khandala (Maharashtra) Kerala; the Indian pied hornbill forests and mango grooves from the plains to 350 m altitude along the terai from Punjab east to north eastern States, and in Bihar, Orissa, Andhra Pradesh and Madhya Pradesh; and the Malabar pied hornbill (*Anthracoceros coronatus*) of evergreen and moist deciduous forest from the plains to 300 m altitude from Ratnagiri (Maharashtra) east to Orissa and south to Karnataka, Tamil Nadu and Kerala.

Reptiles

The Reptiles of India belong to the following groups; tortoises turtles (Chelonia), crocodiles and gharials (Loricata), lizards (Sauria) and snakes (Serpentes). Exploitation of reptilian resources in India is chiefly for flesh and eggs as food, hide, skin and shell for fancy leather articles and handicrafts, teeth and bones as charms, fat for alleged medicinal properties and venom for preparation of medicinal products. Living snake specimens are also utilized by snakes charmers and there is a great demand from the zoos and menageries.

Chelonia

The flesh to turtle, tortoise and terrapin is widely consumed almost throughout India. They are collected from both inland and marine sources. Indiscriminate capture for trade in their flesh and shell, and collection of eggs from rookeries have resulted in many species becoming rate. Particularly endangered are the leatherback

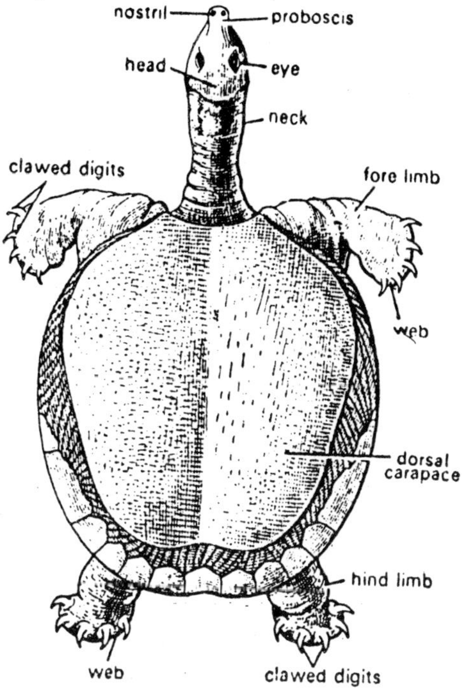

Fig. 10.6. Trionyx gangeticus.

or trunk turtle (*Dermochelys coriacea*), the green sea turtle (*Chelonia mydas*), the loggerhead or olive ridely (*Lapidoeheyls olivacea*) and the hawks bill or tortoiseshell turtle (*Eretmochelys imbricata*). They are all marine forms and found in the coastal areas and estuaries. All the three crocodilians found in India, viz. the esturaine crocodile (*Crocodylus porosus*) found in the estuaries of large rivers and coastal areas, including those of the Bay Islands, the marsh crocodile (*Crocodylus palustris*) occuring in inland swamps and larger wartery-bodies all over India, and the gharial (*Gavialis gangeticus*) found in the big rivers with islands littered with gravel, boulders and pebbles, all lead a precarious life.

Lizards

All the five species of monitor lizards, namely the common Indian monitor (*Varanus bengalensis*) widely distributed throughout India, the desert monitor (*Varanus griseus*) found in the arid and semiarid zones of western, north-western and west-central India, the yellow monitor (*Varanus falvescens*) found in northern India in

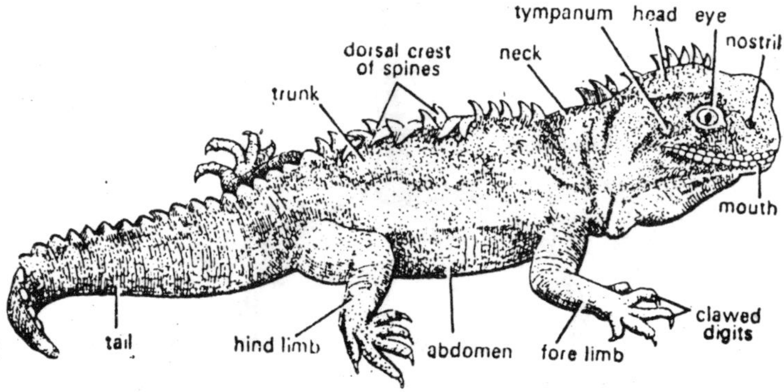

Fig. 10.7. Sphenodon punctatum.

the Ganga Valley, the water monitor (*Varanus salvator*) found in the estuarine and freshwater swampy areas of eastern and south-eastern part of the country, and the clouded monitor (*Varanus nebulosus*) found in eastern India and the Bay Islands are greatly threatened.

Snakes

Of the snakes, the Indian or rock python (*Python molurus*) found in the dense forested tracts throughout India and the reticulated

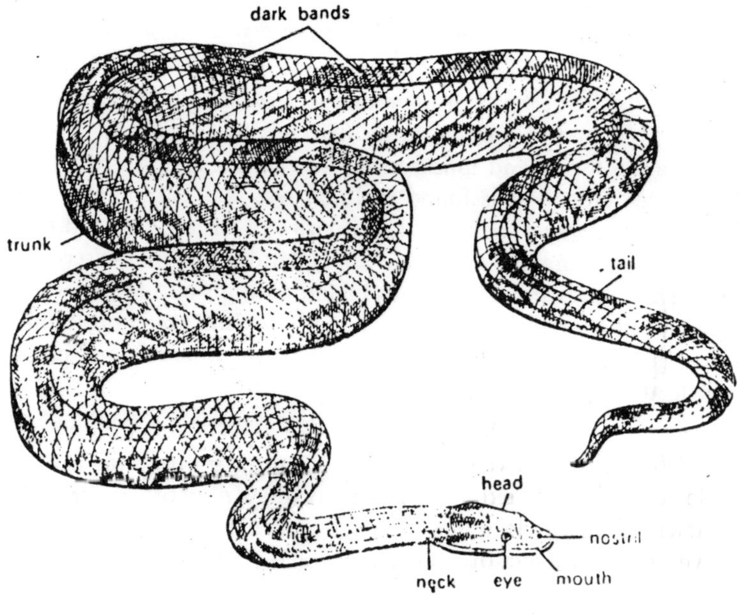

Fig. 10.8. Python molurus.

python (*Python reticulata*) restricted to tropical and subtropical forests of eastern and north-eastern India and in the Bay Islands are largely persecuted for their ornamental skins.

Amphibians

Of the amphibians, the Himalayan Newt or Indian Salamander (*Tylototriton verrucosus*) is of much zoological interest and is found in eastern and north-eastern India. It is restricted to some small pockets in the Darjeeling District of West Bengal, Sikkim, Arunachal Pradesh and Manipur. Habitat destruction and aquatic pollution, it is feared, may greatly endanger its survival. The viviparous toad (*Nectophryne tuberculosa*) is a very rare amphibian from the Malabar rain forest tracts.

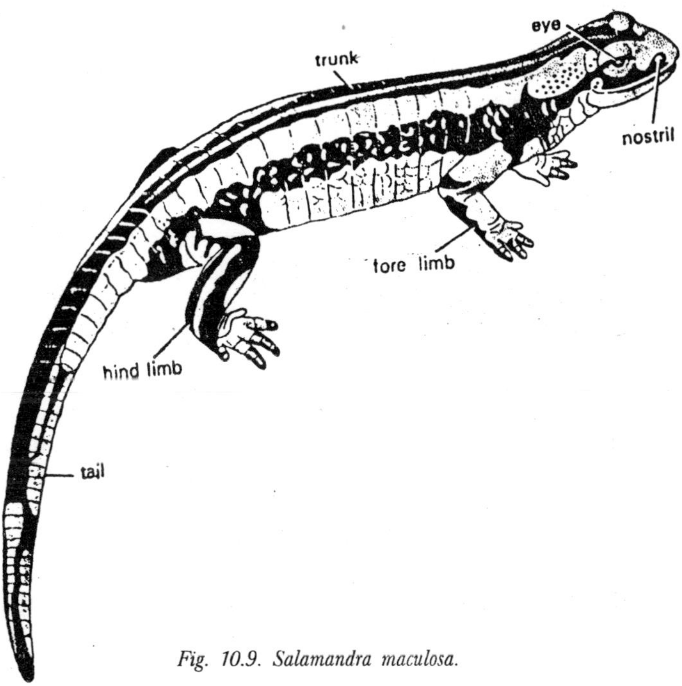

Fig. 10.9. Salamandra maculosa.

Wildlife Conservation

Wildlife Conservation includes all human efforts to perserve wild animals from extinction. It involves the protection and wise management of wild species and their environment. Some species have become extinct due to natural causes, but the greatest danger to wildlife results from human activities. Thus, we ourselves have created this need for wildlife conservation. The progress of man throughout has been beneficial for the human race but it is the wildlife that has suffered through the years.

Invention of sophisticated weapons, industrialization, urbanization, ever increasing human population have been some of the major causes for the dwindle of our once rich resource. Hunting, clearing of forests, draining of swamps and damming of rivers for irrigation and industry this is what we appraise of man's progress. These activities have vastly reduced the natural habitats of our wildlife and many species are endangered or nearly extinct. Man's indifference and ignorance has led him to believe that he is independent of natural laws and of natural environments and forgets that he himself is a biological being and is directly and/or indirectly dependent on the natural resources available to him. He exploits, destroys and reduces the renewable natural resources, eradicates wildlife and leads his own race to levels leading to catastrophe. All this can be put to an end by careful long-range ecological planning.

Importance of Conservation

Alongwith the topic of conservation arises a question–Why should be care about disappearing animals? The answer to such a question can be attributed to many reasons. Some of them are as follows:

Survival value

Every species plays an important role in maintaining an ecological balance among the living systems of the earth. These systems must continue to function if life is to survive. Loss of any species threatenes the survival of several species inclusive of man. Man's own survival depends on his willingness and ability to co-operate with the living environment. Human beings turn to nature and wildlife for spiritual, psychological, physical and cultural benefits.

Scientific value

The study of wildlife provides valuable knowledge about life processes which has helped scientists understand the functioning and behaviour of the human body. Scientists have also gained medical knowledge and discovered important medical products by studying wildlife. By studying the effects of environmental pollution on wildlife it is possible to learn how pollution affects human life.

Economic value

Wild species of animal provide meat for food and skin for fur. The financial value of wild species is of major importance in the economics of many nations.

Beauty

Wildlife includes a diversity of species that contributes in a special way to the beauty of nature. Such beauty enriches life and increase the enjoyment of camping and outdoor recreation. On all continents wildlife is becoming an increasingly important recreational asset and tourist attraction. They maintain the health of the living landscape and productivity of the ecosystem. Wildlife represents the result of many years of evolution and constitute and environmental heritage to the past, a world which once surrounded our ancestors.

Moreover, why should man deny that animals as species have the right to live and coexist with ourselves? We are responsible to the coming generation to leave as an inheritance an upspoiled world with intact living beauty and richness. Apart from the aesthetical ethical reasons, there are also educational arguments to preserve wildlife.

Reasons for Dwindle

Man in his zeal for progress fails to realize that his own existence is being jeopardised. "Progress", as he terms it, has made him contaminate air; pollute streams, lakes and oceans, drain swamps and wetlands, lowering the water table; rape prairies and forests; spray pesticides over water and land; kill animals excessively for immediate profit, or on false assumptions that they are noxious or for fashionable "sport". Man alone, though often in combination with his livestock is destroying the natural environment. "Civilization", as we apply the term, is not just cities industrial developments, miles of highways, enormous airfields, and numerous other technoloigcal conveniences. By destroying wildlife for civilization and culture, we are only negating the meaning of the words. If he does have to change habits, then he must be able to manage them properly in order to maintain a rich and varied wildlife. Man has to think of the dangerous trop he is setting up for himself human over population. It is man's ability to think that formulated the idea of conservation is applied ecology and may also be regarded as bio-economy.

In other words, conservation is defined as sound preservation, management and sensible use of the available renewave natural resources. This decision of conservation is in our hands and man seems to be going in the right direction regarding this decision that holds the key to the future of our most valuable natural resource. Governments of several nations have passed laws and set aside national parks, sanctuaries and other wildlife reserves in an effort to save the dwindling wildlife. Such efforts have saved several species from becoming extinct. However, these efforts hae to increase several fold, if not for a complete recovery, at least about 3/4 of the species threatened with extinction can be redeemed.

Classification of Scarce Wildlife

There are three major categories of wildlife that are facing extinction.

Endangered species

They face the most severe threat of extinction and require direct human protection for survival.

Threatened species

They are abundant in certain areas but face serious danger nevertheless. Dangers being caused due to unfavourable environmental conditions, extensive hunting, fishing, trapping or even collecting by hobbyists.

Rare species

They have small populations. They live in protected environments and their numbers are not decreasing.

Methods of Wildlife Conservation

This depends on the source of danger to the threatened species. In many cases, wildlife can be helped by ensuring that their environment is favourable with regards to availablity of food, water and shelter. This method is called *Habitate Management.* It involves such action as soil conservation, good forestry practices and wildlife management. Several species have been threatened with extinction due to destruction of habitat by human activities like deforestation for farming, pollution of air, water, land, plants and animals etc.

People must set aside separate areas like sanctuaries, wildlife reserves, parks etc. In which wildlife can survive. Hunting is another major cause for the dwindle in the wildlife and this can be prevented by enforcement of strict laws that forbid or regualte such killings.

Excessive increase in the number of one species in a given area threatens not only its own endurance but also that of the other species by consuming major proportion of the available food. Such numbers must be reduced by controlled hunting or by restoring its natural enemies where they have become scarce. Sometimes species find it extremely difficult to survive in their own once favourable environment. Such species can be protected by breeding in capacity and releasing in a protected area that satisfies the conditions favourable for survival.

A species threatened with disease can be protected by sanitation measures in a habitat. The success of wildlife conservation depends on the knowledge of the *Ecology* of the species which involves on understanding of the way of life of the species and its relationship with its biotic and abiotic environment.

History of Wildlife Conservation

The story of the struggle for wildlife conservation is served descades old. If all started with few foresighted men who understood the role of wild animals in living landscapes and the necessity of preserving the threatened species. Encouraged by the pioneering efforts of these men several national and international organizations were formed for conservation of the most valuable of our natural resources. The first of such organizations was formed by a group of Omitholoigsts in 1922, the International Council for Bird Preservation (ICBP). The first post-War organization to be formed under the sponsorship of UNESCO was the International Union For Conservation of Nature and Natural Resources (IUCN) in 1948.

In 1961 the IUCN formed the first fundraising organization, the International Wildfowl Research Bureau (IWRB). There are several other organizations like the Food and Agricultural Organization (FAO) of the United Nations the Smithosian Institute and the Conservation Foundation both in Washington DC and the Fauna Preservation Society in London. The Zoologists Gessellschaft in Frankfurt, West Germany, has raised important funds for conservation, particularly for Africa. The IUCN though its Survival Service Commission (SSC), is dealing specifically with animals threatened with extinction. Their function is to collect data on endangered species throughout the world and to initiate action to prevent the extinction of these species.

The SSC investigates the status and ecology of a species and advises governments and organizations. Such advice is often continued with direct action financed by WWF. The SSC has long maintained a list of threatened mammals, birds, reptiles amphibians, fishes etc., called "The Red Data Book." This conservation e. ɔansion has produced positive results in many parts of the world, Inspite of such efforts, the future remains uncertain for the world's wildlife. This uncertainity exists and intensifies with the increase in human population, destruction of wildlife habitats and spread of environmental pollution.

Wildlife Conservation in India

The Indian Wildlife possesses a rich and diverse fauna and is unique in having immense natural beauty in its different ecosystems. The fauna includes about 123 families of terrestrial vertebates. An estimate puts the number of 400 mammals, 1200 species of birds, 350 species of reptiles and more than 29,70,000 species of insects. The animals that make up the Indian fauna are Elephant, Rhinoceros, Musk deer (Kastura), Barking deer (Kotra), Spotted deer (Cheetal), Hong deer (Hoghiran), Mouse deer (Indian chevrotain), Swarp deer (Bara singha), Dancing deer (Sambur) Thamin, Kashmir stag. Four-horned antelope (Chausingha) Indian Gazelle (Chinkara), Blue bull (Nilgai), Bison, Wild buffalo, Himalayan ibex or wild goat (Capra siberica), Wild boar, Wild ass, Nilgiri tahr, Lion, Tiger, Leopard, Striped hyena, Nilgiri langur (*Ceropithecus John*), Lion tailed macaque (*Macaa silenus*), Rhesus money (*Maaca mulatha*), Hanuman monkey (*macca mulatha*). Hanuman monkey (*Semnopithecus entellus*), Indian giant squirrel (*Ratufa indica*), Procupine, Pangolin, Peafowl (*Pavo cristatus*), Junglefowl, Patridge, Quail Great Indian bustard (*Choriotes nigriceps*). Duck pigeon, Sand grouse, Stroks and Egrets, Pelicans, Eagle, crane, Owl, Hornbill, Crocodiles, Gharials, Lizards (Uromastic) and about 216 species of snakes.

Endangered Species of India

The IUCN maintains a Red list is threatened species. This list is updated annually. A 1986 list contained the following species that were threatened with extinction. These species face the most serious threat of extinction and require direct human protection for survival. In order to protect these species from extinction we have to know about their habits and habitats. Once we are through with such facts, preservation of the endangered species can be carried out successfully. The above table shows the list of endangered species in India and a detailed description of each animals is as follows:

Order	*Family*	*Zoological Name*	*Common Name*	*Status*	*Habitat*
1	2	3	4	5	6
	Class Mammalia				
Primates	Cercopithecidae	Macaea silenus	Lion tailed macaque	E	Southern Western Ghats
Lagomorpha	Leporidae	Caprolagus hispidus	Hispid Hair	E	Assam
Cetacea	Platanistidae	Platanistaindi	Indus dolphin	E	India
Carnivora	Viverridae	Viverra Megaspila civettina	Malabar spotted civet	E	Southern Western Ghats
	Felidae	Panthera leopersica	Asiatic lion	E	Gir Forest
		Panthera tigris	Tiger	E	India
		Panthera uncia	Snow leopard	E	Himalayas
Proboscidae	Elephantidae	Elephas maximus	Asiatic elephant	E	South NE India
Perissodactyla	Equidae	Equus henionus khur	Indian wild ass	E	Gujarat
	Rhinoceridae	Rhinoceros unicornis	Great Indian Rhino	E	Assam North Bengal
Artiodactyla	Suidae	Sus sulvanius	Pygmy hog	E	NE India
	Cervidae	Cervus duvaucelli	Hungal	E	Terai Kashmir
		Cervus elaphus	Swamp deer	E	Central India

		Hungal Cervus eldi eldi	Manipur deer	E	Manipur
	Bovidae	Bubalis bubalis	Wild asiatic Water buffalo	E	Assam Bastar
	Class Aves				
Galliformes	Phasianidae	Tragopan caboti	Cabot's Tragopan	E	NE India
		Tragopan melano cephalus	Western Tragopan	E	Western Himalayas
Gruiformass	Otididae	Ardeotis nigriceps	Great Indian Bustard	E	Rajasthan MP
	Class Reptilia				
Testudines	Crocodylidae	Crocodylus porosus	Estuarine crocodile	E	East coast
	Garialidae	Garialis gangeticus	Gharial	E	N Indian river
	Class Insecta				
Anoplura	Haematopmidae	Haematopinus oliveri	Pygmy hog sucking louse	E	Assam

E–Endangered species.

Panthera Tigris

Common Name : Tiger

Local Name : Hindi–Bagh, Central India–Nahar, Tamil-Poolee, Malayalam-Koduwa.

Size : Average length 9 ft. –9.6 ft. (275-290 cm). Female-8 ft. 6 in (260 cms.) Average weight –400-500 lbs (180-230 kgs). Female–100 lbs less (45 kgs.)

Distinctive characters : Tigers have a rich-coloured, well stripped short coat. Individual variations are great.

Distribution : They are found throughout India from the Himalayas to cape Comerin except in the deserts of Punjab, Sind, and Cutch. Also found in Burma.

Habits : Tigers are found at 10,000 ft. in the Himalayas, humid evergreen forests, in dry open jungle, grassy swamps of Terai, in Sunderbans leading an amphibious life, on trees, in mud and in water. Food, shade and shelter are essential for their survival. Tigers hunt from sunset to dawn, when it is not too hot. Its prey comprises of elephant (female and young), nilgai, deer, wildpig, bears, porcupine, gaur, buffalo, panthers, other tigers and when extremely hungry feeds on fowl or fish. When game is not available, tigers attack cattle and even man. They climb trees and are excellent swimmers. Though they are heavy, they are athletic.

Mating is usually after the rains. Young ones are born between February and May. Male-female relationship ends when young are born: Gestation period is about 15-16 weeks. Tigresses give birth to 2 to 3 cubs per litter. The cubs begin to hunt when they are 6 months old and they stay with the tigress for 2 years. Sexual maturity is attained at the age of 3 years in the case of the tigress and 4 years in the case of the male.

Life spain of a tiger is about 20 years. Hunting had greatly reduced the number in India. In 1970, a project was set in motion to prevent the degradation of the tiger population. This project was called "project Tiger". Nine sanctuaries were set aside as Tiger reserves for rehabilitation of the species. At the beginning of the project the number of tigers in India was less than 2000.

Macca Silenus

Common Name : Lion Tailed Macaque.

Local Name : Hind Siah Bundar, Malayalam–Nellamanthi, Tamil–Karungkorungoo, Arakkan.

Size : Head and Body 20-24 inches. Tail- 10-15 inches (25-38 cms.) Females slightly smaller.

Distribution : Western Ghats from North Kamara, southwards to Kerala and Kanyakumari district, Tamilnadu.

Distinctive Characters : Great mane of long dark-grey or brownish grey hair from temple and cheek, Glossy black coat.

Habits : In habitats lonelier evergreen tropical forest belt between 2000-3500 ft. (610-1070 m). They are dark coloured, shy and seclusive in habit. They are seldom seen Macaques are gregarious. They live in herds of 12-20 animals or more. The call of the male resembles that of the human voice– "COYEH" or the "COO" of a pigeon. Their movement through the treetops are slow and deliberate. Newborns are seen in September in South India.

Destruction of habitat and poaching have made this species an endangered species. Trees are the means of transport and form the habitat of this species. Hence afforestation and prevention of poaching are the methods to be adapted for saving this species.

Panthera Leo Persica

Common Name : Asiatic Lion.

Local Names : Hindi–Sher, Gujarat–Untia bagh Malayalam–Simham, Tamil–Singham.

Distinctive Characters : Mane is lesser than of the African Lion, cost is fuller and longer tassel of hair at the end of its tail, more, pronounced tuft of hair at the elbow joints and fuller fringe of hairs on its belly. The mane is light to dark, rarely, black. Cubs are spotted or striped.

Distribution : One found in the whole of Northwest and Central India as far south as Namada. Now lions are restricted to the Gir forest Kathiawar, Gujarat.

Habits : Gir forest in Junagadh district covers about 500 square miles of rugged country and is composed of mainly stunted teak trees, palas, junbal, ber, patches of bamboo with an undergrowth of thorny shrubs and bushes.

Lions hunt at dusk and rest during the day. Their roars are heard throughout the night. The prey on game and cattle. Lions do not have a particular breeding season. They are known to mate during October and November. Young ones are born during January and February. Gestation period is about 116 days. The male stays with the family and helps to defend the young and to get food for the young. Young ones are produced at intervals of at least 18 months to 2 years. A lioness gives birth to its first little when it is 21/2-3 years old. There may be 2, 3, or even upto 5 cubs per litter. In 1978, the number of lions in the Gir forest was less than 200.

Panthera Uncia

Common Name : Snow Leopard

Local Names : Simla–Barhal he, Kunawar–Thurwagh, Kumaon–Burhel haye.

Size : Smaller than a panther with a relatively longer tail. Head and Body -3ft–3 in-3 ft 8 in (100-110 cm). Tail - 3ft (90 in).

Distinctive Characters : The snow leopard has a short muzzle, high forehead with a vertical chin. The colour of the coat is soft grey palling to pure white on the underside. Grey is sometimes tinged with buff. Spots are unbroken and distinct on the head, nape and lower parts of the limbs. On the body they break up into larger, paler rosettes. These rosettes are less pronounced in the luxuriant winter coat. The undersides are pale white except for a few black biotehes and fur. Newly born cubs are darker than the adults.

Distribution : In India, the range extends along the whole of the Himalayan chain from Kashmir to Bhutan northwards into Tibet, Central Asia and Atlas.

Habits : Not much is known about the snow leopard. Its home is the higher altitudes of Himalayas at heights of 12,000-13,000 ft. (3660-3965 m) above sea-level, above the treeline where the rocks and cliffs form the mountain sides. Lying by day, snow leopards hunt at night preying on wild sheep and goats, musk deer, hares, marmots and other rodents, or even larger birds. In summer, when the upland pastures are open to grazing, they follows their prey to these higher reaches and finding opportunity feed on domestic animals like goats, sheep or ponies. They establish territory near human settlements and feed on domestic animals. In winter, they move down with the animals. In winter, they move down with the animals to about 600 ft. (1830 m). The gestation period is about 3 months. Snow leopards give from 2-4 young per litter.

Uncias are persistently hunted for their valuable fur. Their soft colour and luxuriant beauty is scarcely rivalled.

Caprolagus Hispidus

Common Name–Hispid Hare of Assam Rabbit. Not much is known about this species except that they are the same size of the hare with a shorter, dark-brown tail. The fur is coarse and is made of dark-brown. Bristies which are paler on the under-surface of the body. The outer side of the ear is brown. The eyes are small. Hindlegs are short and stout. Their claws are long and teeth are large. They feed on bark and roots. Usually seen in Terai, Dooars and Himalayan foothills. They are very elusive.

Platanista Indi

Common Name : Indus Dolphin.

The indus dolphin was considered as separate species. Now considered conspecific with Platanista gangetic. They are found in the Ganges, Brahmaputra, Indus and other large tributories to the bases of hills. They are seen within tidal limits, but do not enter the sea. Its body oil is used for burning lamps and for various other purposes. Not much is known about this species.

Viverra Megaspila

Common Name : Malabar Large Spotted Civet.

This species was once very common in the coastal districts of Malabar and Travancore. It lied in the wooded plains and in the adjoining hill slopes. At present they are scarcely seen and appears to be nearly extinct. Nothing else is known about this species.

Elephas Maximus

Common Name : Asian Elephant

Local Names : Sanskrit–Hasti, Gaja, Hindi–Hathi (male, Hathni) (Female), Marathi–Hatti Tamil–Anai, Malayalam–Ana, Kannada–Ane.

Size : Average height-Adult male-9 ft. (2.75m). Female–A foot lower (30 cms).

Distinctive Characters : The Asian elephant is smaller than the African elephant. While the ears of the African elephant or enormous, that of the Asian elephant are smaller. The major distinctive feature is that of the hollow back of the African elephant to that of the rounded back in the Asian counterpart. Generally only the males have tusks, while females have a protrusion of a few inches or nothing at all.

Distribution : In India, the Asian elephant is found in the Western Ghats, southwards from Mysore, Orissa, Bihar, Himalayan range, almost all over India.

Habits : The Asian elephant is found in areas covered with forests. They are extremely adaptable and live in steamy, humid jungle or in cool elevated forests. They live in herds of 5-60 animals (usually single families). When fodder is scarce herds break up into smaller parties. Big tuskers feed away from the herd. Bulls lead a solitary life and joins a herd when he feels the urge to mate. Elephants sleep when it is hot and feed in the mornings and evenings and during nights invade open forests and raid crops. They raid and retire after midnight. They feed on various kinds of grasses and leaves, stem and leaves of wild bamboo and plantain, all species of crops, barks of some kinds of trees, A full grown elephant eats about 600-700 lbs (270-320 kgs). of fodder a day. The exact breeding season is uncertain, bat is estimated at the beginning of the rains. Young appear in Autumn. The gestation period is about 20 months. Elephants give birth to only one calf at a time. Elephants have poor sight but their sense of hearing and small are well developed. Hunting and poaching for ivory has very greatly decreased the number of elephants in India.

Equus Hemionus Khur

Common Name : Indian Wild Ass.

Local Names : Hindi–Gharkar.

Size : Height at shoulder-3 ft. 8 in (110 cms) to 4 ft. (120 cms).

Distinctive Characters : The colour of the coat varies from reddish grey to fawn or pale chestnut. The erect dark brown mane is continued as a dark brown stripe extending along the back to the root of the tail, the lower being white.

Distribution : The Indian Wild Ass is found in the deserts of the Rann of Cutch and Peninsular India. The present population is less than 2000.

Habits : They roam in groups of 10-20 or in 2 or 3's or as individuals in the scorching sun. They are capable of remaining in the open for hours on end and run for a great distance at a speed of 30-32 miles per hour. The Indian wild ass mate in the months of August, September. The gestation period is about 11 months. Young ones appear in the months of July, August, September. They are called *Foals*. They are readily tamed when young. The grow vicious and untameable animals. The present population in India is below 2000.

Rhinoceros Unicornis

Common Name : Great Indian Rhinoceros.

Local Names : Hindi-Gainda, Gargadan: Marathi-Genda.

Size : Average height of male-5 ft. 8 in (170 cms.) with a girth of 11 ft. (335 cms) behind the wither.

Distribution : The Great Indian Rhinoceros was formerly extensively distributed throughout the Gangetic plain, but is now restricted to the Dooars and parts of West-Bengal and Assam.

Distinctive Characters : The skin is divided into shields by heavy folds before and behind the shoulder and in front of the things. The fold is no continued across the back which is a distinctive character in the Indian species. Another distinctive character is that the horn is smaller in the Indian Rhinoceros.

Habits : Rhinos prefer swamps and grasslands but are also found in wood jungles and ravines of low hills. They lead a solitary life. Rhinos feed on grass. They breed through out the year. Their Gestation period is 16 months. The young ones are born as heavy as 60 kgs. and are about 105 cms in length. The female Rhino acquires sexual maturity at the age of 4 years while the male takes 7 years to attain sexual maturity.

The Indian Rhinoceros is a fast diminishing species. They are hunted for sport and their skin is used for making leather products. Their horn is supposed to have medicinal value. In India, they are found in majority only in the Kaziranga Sanctuary in Sibsagar/ Nowgong district in Assam. The Bengal Rhinoceros Preservation Act of 1932 and the Assam Rhinoceros Preservation Act of 1954 were fruitful in protecting the animal and by the end of 1978 the population of the Rhino in the Kaziranga Sanctuary was estimated to be around 800.

Sus Sulvanius

Common Name : Pygmy Hog.

The Pygmy hog is scarcely 10 inches high (25 cms). They inhabit the forests of the Himalayas in Sikkim, Nepal, Bhutan and Assam. Their habits are similar to that of the wild boar. They live in herbs of 5-20 in number. They are nocturnal by nature and rarely seen. Hence nothing much is known about this species.

Cervus Duvanceli

Common Name : Swamp Deer.

Local Names : Hindi–Barasingha, Maha, Central India–Goinjak (male), Gaoni (female), Sal samar, Nerwari.

Size : Average height 54 in (135 cms) at the shoulder, Average weight 370-400 lbs (170-180 kgs). Average length of the horn–30 in. (75 cms) with at girth of 5 in. (13 cms) at mid-beam.

Distinctive Characters : The coat is almost woolly and has colour shades from brown to yellowish brown. Stags are maned and darker in colour. Young ones are spotted. The size and shape of the antlers vary.

Distribution : Swamp deers are found in Terai, Uttar Pradesh, Assam, Sunderbans, Madhya Pradesh.

Habits : They live on marsh land, seldom out of water. They are also found on grassy maidans. They are highly gregarious by nature and less nocturnal. Swamp deers feed till late morning and in the evenings, taking rest during most of the hot day. Their eyesight and hearing are moderate, but their sense of smell is well developed. They are an endangered species and strict protection is necessary to prevent their total extinction. The total number of Central India is less than 200.

Cervus Elephas Manglu

Common Name : Hangul.

Local Names : Kashmir–Hangal, Hanglu (Male), Minyamar (Female) Hindi–Barasingha.

Size : Average height–40-50 inches (120-125 cms) at withers. Average weight-400 lbs (180 kgs). Average length of horn-40 inches (100 cms).

Distinctive Characters : The coat is light to dark brown fading to a dinghy white on the lips, chin, underparts and buttocks. The white rump patch is divided by broad median stripe extending to the base of the tail. Fawns are spotted.

Distribution : Northern side of the Valley of Kashmir and some adjacent valleys.

Habits : The Hungul is basically a forest animal and leads a solitary life or live in small parties. They roam around in search of food. They give birth to their young ones in late May. They spend their time in summer at heights of 900-12,000 ft. (2750-3680m). They shed their antlers before they migrate and grow new ones. This is a very regular feature among this species. In winter they descend to lower levels and feed on fresh grass. Another of the endangered species that requires direct human protection.

Cervus Eldi Eldi

Common Name : Manipur Brownanlered Deer.

Local Names : Manipur–Sanga, Burma–Tamin.

Size : Average height 4 ft. (120 cms) at withers, hinds are smaller. Average length of antler–42 inches (107 cm) Average length of head–over 35 inches (90 cms).

Distinctive Characters : The coat is coarse and sparse. The stags have a dark brown, or black coat. Hinds are light brown or fawn. Young ones are spotted. Antlers are almost erect.

Distribution : Manipur.

Habits : The Manipur Deer avoid hills and heavy forests. They prefer open scrub jungle or flat undulating land between rivers and hill ranges. They live in small herds. Their main food comprises of grass on which they feed in the mornings and evenings. Their speed and their sharp eyesight protect them from their predators. Young ones come in the month of October, Gestation period is about 239-256 days. The present number is only about 100 most of which are restricted to Keibal Lanjao Sanctuary on the shore of Logtak Lake in Manipur. This is the reason that they were given the name Manipur Deer.

Bubalis Bubalis

Common Name : Wild Asiatic Water Buffalo.

Local Names : Hindi–Arna (Male), Arni (female), Marathi – Jungli mhais, Kannada–Kadu Kano Kartee.

Size : Average height of the Bull-5 ft. 6 in. (170 cms) at shoulder to 6 ft. Average weight–2000 lbs (900 kgs) or more.

Distinctive Characters : The wild buffalo is sleeker, heavier had more robust-looking than their tame counterpart. Their body is slaty black in colour. The newborn are light coloured, almost yellow in colour. Their horns are curved or spread outwards. The horns may be flat, triangular in cross-section, the horn of the cows are less massive and vary less in shape.

Distribution : They are found in grass jungles of Terai and the plains of Brahmaputra in Assam, Orissa and Bastar District of Madhya Pradesh.

Habits : Tall grass jungles with food, shelter and water to lie in, form the natural habitat of the wild buffalo. They live in small herds. They feed chiefly on grass. Feeding is done only in the mornings and evenings and sometimes in the night. During the day they lie down in the shade. Wild Buffaloes are bold and most savage by nature. They invade cultivation in search of food. Bulls. may attack without warning or provocation. A cow with a new born calf is very dangerous. The chief enemies of this species are man and tiger. Their sense of smell is well developed and they posses good hearing. They have moderate eyesight. Calves appear in the months of March, April or May. Wild Bullaloes sometimes mate with domestic ones.

Details about only a few species have been mentioned as information about the rest has been pressed to find. The above facts have been mentioned only as an example as to how one can proceed with collecting data about an endangered species for the purpose of conservation.

Modes of Wildlife Conservation in India

In India, several laws have been passed and sanctuaries and national parks have been established for the protection of the dwindling wildlife. India was probably the first country to enact a Wild Life Protection Act. The Wild Birds and Animals Protection Act was passed in 1887 and repeated in 1912. For game protection in the states, in 1927, the Forest Act XVI was enacted. Indian Board for Wild Life was established in 1952 and this was followed by setting up of Wild Life Boards in different States in India. In 1972, a new Wild Life Protection Act was passed. Under this Act, possession, trapping, shooting of wild animals alive or dead; serving their meat in eating houses; their transport and export are all controlled and watched by special staff (Chief Wild Life Warden and authorized officers). This act prevents hunting of female and young ones. Under this Act the threatened species are absolutely protected and the rest afforded graded protection according to their state of population size.

In India, nearly 200 sanctuaries and national parks have been established for wildlife management. Sanctuaries are places where the killing and capturing of any animal is prohibited except under orders of the authorities concerned. National Parks are set up for preserving flora, fauna, landscapes and historic objects of an area. Some well known wildlife sanctuaries and national parks of India are:

Kaziranga Wild Life Sanctuary

It was established in 1926 in the district of Sibsagar, subdivision Jorhat (Assam) on the southern bank of the Brahmaputra river. It covers an area of 430 sq. kms of the forest grasslands and swamps and supports a fauna of 700 Rhinoceros besides elephant, wild buffalo, bison, tiger, leopard, sloth bear, sambhar, swamp deer, hog deer, barking deer wild boar, gibbon and birds like pelican, stork and ring-tailed fishing eagles.

Mans Wild Life Sanctuary and Tiger Reserve

It is located in the district of Kamrup in Assam, convering an area of 540 sq. kms. and is situated at an altitude of 80 metres. The River Manas runs through this sanctuary. It contains the following wild animals: tiger, panther, wild dog, wild boar, rhinoceros, gaur, wild buffalo, sambhar, swamp deer and golden langur.

Jaldapara Wild Life Sanctuary

It is situated in the Jalpaiguri district of West Bengal and covers an are of 65 sq. kms of grassland. Wild life fauna includes animals like rhinoceros gaur, elephant, tiger, leopard, deer and a variety of birds and reptiles.

Palamau National Park

It is situated in the Daltongunj district of Bihar and has an area of 345 sq. kms. The fauna of this include tiger, panther, sloth bear, elephant, chital, gaur, nilgari, tiger, chinkara, chowsingha and mouse deer. The Flora is thick tropical forests.

Hazaribagh National Park

This national park was established in 1954 in Bihar. It has an area of 184 sq. km of thick tropical forests. The typical fauna of this park includes wild boar, sambhar, nilgai, tiger, leopard, sloth bear, hyena and gaur.

Simplipal National Park

It is situated in the district of Mayur Bhanj in Orissa and has an area of 2750 sq. kms. It is covered with dense sal forests and has been chosen for Project Tiger. Its typical fauna includes tiger, elephant, deer, pea fowl, talking mynas, chital sambhar, panther, gaur, hyena, and sloth bear.

Chilka Lake

This largest inland lake has an area of 1000 sq. kms. and is about 1000 kms. from Bhubneshwar (Orissa). Its fauna includes waterfowl, duck, cranes, ospreys, golden plover, sand-piper, stone curlews, flamingoes, etc.

Kolameru Bird Sanctuary

This is a small bird sactuary near Tadepallegudam in Andhra state. It is a breeding place for pelicans and many marine birds visit this place.

Vedanthangal Bird Sanctuary

This small-sized but very old sanctuary extends over a lake of about 0.30 sq. kms. 85 kms from Madras. Many migratory birds regularly visit temporary lake. The regular visitors of this lake are birds like spoon bills, open billed storks, egrets, ibis, cormorant darter, grey heron, pelicans, snipes and dab chicks.

Guindy National Park

It is situated near Madras city and has mainly chitals and black bucks. A few albinos of black buck are also found here.

Point Calimer Wild Life Sanctuary

This is situated at the southern tip of the Thanjavur district of Tamil Nadu abutting the Palk Strait. Its back water and lagoon is visited by flamingoes and pelicans. It is close to the Vedaranyam forests which has a fauna of numerous blacks bucks chitals and wild boars.

Mundanthural Sanctuary

It is situated in the Tirunelveli district of Tamil Nadu and was established in 1964. It covers an area of 520 sq. kms. The Tamaraparani river flows through the evergreen forests of the sanctuary. Panther, tiger, sambhar and chital form the major fauna of this sanctuary.

Periyar Wild Life Sanctuary

Situated in the state of Kerala, this sanctuary covers an area of 777 sq. kms. It was established in 1949 around the artificial lake which arose behind the dam built across the Periyar river in 1900. This sanctuary of great scenic beauty supports a fauna of wild elephants, gaurs, leopards, sloth bear, sambhars, barking deer, wild dogs, wild boars, black Nilgiri langur and water birds like grey hornbills, egrets.

Mudumalai Wild Life Sanctuary

This wildlife sanctuary was established in 1940 in North Western part of Nilgiris in Tamil Nadu. It is known for its rich forests and diversity of fauna that includes wild elephants, gaur, sambhar, chital, barking deer, mouse deer four-horned antelope, tiger, panther, bonnet monkey, langur, giant squirel, flying squirrel, wild dog, jackal, wild cat, sloth bear, porcupine, pangolin, flying-lizard, monitorlizard, rat snake, python and various birds.

Ranganthito Bird Sanctuary

It covers an area of 166 sq. kms and includes a series of islands in the Cauvery river 15 kms of the Bangalore-Mysore road, near Shrirangapatnam. Its avial fauna includes open bill strok, white ibis, erget, spoon bill, wild duck, pea fowl, night heron etc.

Bandipur Wild Life Sanctuary

This sanctuary was established in 1941 by the then ruler of Mysore (Karnataka State). It is situated 80 kms south of Mysore city enroute to Ootacamund. It has an area of 874 sq kms and is at an altitude of 1454.4 metres. Its forest is very thick and has plenty of rain fall. Its wild life fauna includes plenty of gaurs and animals like elephant, leopard, sloth, bear, wild dog, chital, panther, barking deer, porcupine and lungur.

Cotigo Wild Life Sanctuary

It is located in South Goa and has an area of 105 sq. kms. It has wet evergreen forest of Elve and Berla trees and supports rich fauna of the following animals, gaur, sambhar, chital, hog-deer, barking deer wild boar, hyaena, panther, leopard, jackal, otter, porcupine, and birds like parakeet lorokeet, woodpecker, kingfisher, bulbul, jungle fowl, egret etc.

Bhagwan Mahadev Wild Life Sanctuary

It is located in North Goa and has an area of 240 sq kms and supports fauna almost similar to Cotigao Wild Life Sanctuary.

Sesan Gir

This famous wild life sanctuary for the Asiatic lion is situated in Gujarat state, 468 kms from Ahmedabad and 43 kms from Veraval. It has an area of 1295 km sq of semiarid country with patches of thorm scrub and deciduous trees. Its fauna includes Asiatic lion, spotted deer, blue bull (nilgai), fur-horned antelope, chinkara, striped hyena, wild boar, porcupine, langur, python, crocodiles and birds like green pigeon, partridge, rockgrouse etc.

Kanha National Park

This national park was established in 1955 in former Banjar Valley Reserve (Madhya Pradesh). This park has an area of 939.94 sq kms. and includes hilly terrain and streams. It is 175 kms away from Jabalpur and has forests of sal trees. Its typical fauna includes animals like tiger, chital, panther, sambhar, black buck and barasingha.

Tandoba National Park

It is located in Chandrapur district (Maharashtra) and has an area of 116 sq. kms. Its fauna include tiger, sambhar, sloth bear, bison, chital, chinkara, barking deer, blue bull, four-horned deer, langur, peafowl and few crocodiles.

Sariska

This is one of the most beautiful sanctuaries situated in the state of Rajasthan near Alwar. It has an area of 800 sq. kms. and has dense Dohokara and Solar forests. Its fauna include tiger, leopard, spotted deer, jungle cat, four-horned antelope, langur, porcupine, hedgehog, peafowl etc.

Bharatpur Bird Sanctuary (Koeideo Ghana)

It is located in Bharatpur district in Rajasthan, has an area of 29 sq. kms. and harbours all kinds of indigenous birds like nesting water birds, water side birds and migratory birds. More than 328 varieties of birds including cormorants, spoonbills, white ibis, Indian darters, egrets, painted storks, open billed storks, great black necked stroks, etc. Many migratory birds like ducks, geese, siberian cranes, etc. regularly visit this sanctuary. Drier parts of this marshy sanctuary have animals like spotted deer, black buck, sambhar, blue bull, wild boar and python.

Sultanpur Lake Bird Sanctuary

This small sized (2 sq kms) bird sanctuary is located in Gurgaon district (Haryana) about 30 kms from Delhi. Its typical avian fauna includes crane, sarus, spot-bill, rudyshell and drake.

Shikari Devi Wild Lie Sanctuary

It is located in Mandi district in Himachal Pradesh. It has an area of 213 sq kms and has the following animals, snow leopard, flying fox, black bear, barking deer, musk deer, chakor, partridge, etc.

Bir Motibagh Sanctuary

It is located near Patiala in Punjab. The fauna includes black buck, blue bull, hog deer, hare, jackal and birds like peafowl, partridge, sparrow, boller, myna, parakeet, pigeon, dove, etc.

Dachigam Wild Life Sanctuary

It was established in 1951 in Kashmir, 26 kms away from Srinagar. It has an area of 89 sq kms and has two levels: upper Dachigam at 3692.3 metres altitudes and Lower Dachigam at 1846.2

metres altitude. It mainly preserves hungul or Kashmir stag, musk deer, leopard, black buck bear, brown bear and babboon.

Corbett National Park

It is one of India's most famous wild life sanctuaries and was constituted in 1935 as the first national park in India. It is situated between Nanital and Gharwal districts in Uttar Pradesh. It has an area of 525 sq kms and is located within west-south bend of the river Ramganga. It supports a rich and diverse fauna of the following: tiger, panther, sloth bear, hyanea elephant, blue bull, swamp deer, barking deer, Indian antelope, porcupine birds like bulbul, wood pecker, babber, and bee eater, and reptiles like crocodile, python etc.

Shivpuri Sanctuary

It is located in Madhya Pradesh and is an asylum for tigers.

Annamalai Sanctuary

This sanctuary was established in 1972 in the southern part of Coimbatore district in Tamil Nadu. It has an area of 958 sq. kms and supports rich fauna of animals like elephant, gaur, sambhar, spotted deer, barking deer, nilgai, Nilgiri langur, Nilgiri tahr, lion-tailed macaque, tiger, panther, sloth bear, porcupine and pangolin.

These are some of the major sanctuaries in India. There are several smaller sanctuaries and national parks that house some of the important species. A sanctuary or a national park may be large or small, but planning should be based on a scientific background. To provide the basic needs for the wild life the following steps may be taken as far as possible for conservational purposes.

1. Scientific studies of the sanctuary wildlife particularly of threatened species, by qualified persons, should be encouraged in order to improve the status of the animals even by breeding them in captivity to rehabilitate them in suitable habitats.
2. Concentrated grazing by domestic live-stock is to be prevented as wildlife can never compete with domestic stock under ordinary circumstances and domestic animals may be responsible for transmitting diseases like foot and mouth disease, rinderpest, surra (sleeping sickness), haemorrhagic sepicaemia, arthrax etc.
3. Cultivation near the sanctuary should be avoided as rule as it may have an indirect effect on the wildlife. The chemical control

of certain pests is often desirable for better yield in agriculture, but these pesticides can present ecological hazards for wildlife.

4. The water holes and stream sides should be managed and guarded carefully for the benefit of wildlife. Such critical areas may be of small size, but their importance is disproportionately great because they supply important elements in wildlife ecology for largest areas during hot months of the year.
5. The forest should not be planted with or allowed to grow only a single species of plant. Further, fruit-bearing shrubs and trees should be encouraged where ever possible and the natural openings of the forests should be preserved.
6. Annual burning of the grassland in most of our sanctuaries has become an intergral part of management, since not burning these might bring about a lack of suitable fodder during the hotter months. Controlled burning has been found to increase forage and to preserve organic material in the soil. Such burning may be done in blocks with unburned areas in between, so that areas are burnt in rotation. The burning is to be confined to definite blocks by means of five-breaks. Burning at night is favoured unless it is likely to extend the fire beyond the desired limits.
7. Provision for dust bath for animals should also be made by leaving some exposed soil for their dusting. These spots may be prepared in the form of small mounds elevated a few inches from the forest floor to allow for drainage and to dry quickly in wet months.
8. Some artificial salt-licks are also so be provided in the sanctuaries since the wildlife need them at regular intervals for their normal health. Such salt-licks are necessary for animals especially in areas where natural salted clay is scanty.
9. The forest inside the sanctuary park should be left unexploited as far as possible. It should be an ideal environment for the wild life to get the fullest measure of protection, and in such preserved conditions of life they can multiply freely.
10. If total unexploitation is not possible, the cutting operations of the forest should be in small units, well distributed over the forest. Frequent light cuts are generally better for wild life than infrequent heavy cuts.
11. Poaching needs to be ruthlessly prohibited in the sanctuaries in particular and generally in other areas outside them, where

wildlife still exists. Mobile armed units should be attached to all sanctuaries to petrol regularly the entire sanctuary to stop illegal activities.

Most wildlife could be saved by sufficiently rigid protection of their natural habitat, combined with controlled exploitation of species which are of economic importance. Some wildlife may be driven away from their natural habitats so that their requirements are not in conflict with human interests. The only possible course for maintaining their viable number is in national parks, sanctuaries, game reserves, where they could thrive unharmed. In the case of the most endangered species which have reached a points of extinction in their favourable habitat, their survival may be ensured through the maintenance of breeding stock in zoological parks in suitable parts of the country.

Man, An Endangered Species

"Man is part of the vast web of life and cannot escape the natural consequences of his actions. The sheer power of the population and technological revolutions may make man himself and endangered species in many parts of the world".

– *Stewart Udall*

The present day present Man as the most successful species in the world. The ability to think, a sixth sense, is what places the species of Homo sapiens at the top of the ladder. It is this ability that has assisted Man to become the most dominant form of life on earth. In the game of survival, Man has come out in flying colours, this being possible only by his dependence on the environment around him. Initially, this progress of Man was very slow, but with time it has increased to a dizzying pace, this being made possible by the explosion of scientific knowledge and its application to every phase of modern life.

Man's dream for the future is machine oriented civilization. This coupied with his greed is definitely going to have an adverse effect on his natural environment, soul and spirit. An endangered species is one whose population is decreasing. In man's case it is the ever increasing population that is making him an endangered, species. There is a general belief–more people move technology–which if followed blindly, as is being done, is to flirt with ultimate disaster.

The result of progress and technological advancement is the pollution of water, land, air, removal of forests and other natural resources beyond redemption. All these activities tilt the ecological balance on which all life depends. Each species is a part of a food chain and man is no exemption. Thus preservation of wildlife forms a major decision in the fate of man Man's attitude towards nature only suggests that he is not ready to believe this fact. Unless this attitude changes, all one can decipher is that man is paving roads leading of extinction. President Johnson of the United States of America in an address to the Congress in 1968 observed that–"Conservation's concern now is not only for Man's enjoyment but for Man's survival". Modern Man, if he is to survive has to improve his relationship with nature. If he succeeds–he must, he will–all wildlife too will have a better chance.

The Future

The future of our wildlife looks very bleak. If wildlife is not preserved then this world would be a paradise lost, to put it in better terms, a paradise destroyed. As Man's life becomes more and more urbanized, mechanised and machinised, the habitat for wildlife is being devoured and wild animals are losing their place on our globe. Preservation of habitat is the most important criteria in conservation of wildlife. With the ever increasing human population, the idea of conservation can be scarcely brought about.

It is a habit for nature of eliminate anything that exceeds a limit and Man, the most powerful and highly developed of living creatures, fails to recognize the needs for his own existence in future. Nature and wildlife are of definite economic value and components of an environment of which Man is a part and on which he is directly or indirectly dependent. Man enjoys music, poetry and paintings, all he has to realize is that the flora and fauna are the music; poetry and paintings created by nature. Even more, they form a part of the food chain which is a guarantee for the continuity of life on earth. Very few individuals is this world have time of recreation, leisure or intellectual interests. The rest of them think wildlife conservation is a luxury. Each of them is too busy trying his own problems for him to even consider the idea of conservation.

Starving illiterates have time to think only of ways to earn their next meal and not about conservation, which for them is nonfunctional luxury. Human overpopulation and illiteracy are two of the major problems that bring about the tragic destruction of environment and animal exterminations in the world. For every extinct or endangered species or subspecies of animals, two of the major factors have been the destruction of habitats and excessive cropping by man. Connected with this depletion is the lack of knowledge about the essential needs of a species for survival, its needs for habitat, and its reproductive capacity.

Several organization are painstakingly trying to bring about a successful turn-about regarding the concept of conservation. They educate and advise on the need for conservation of ourdepleting natural resources. All these organization are run by men who are devoted and work their hearts out for saving habitats from destruction and species from extinction throughout the world. They fight against social, economic, political and religious problems in different parts of the world. It all looks like a hopeful sign that the future of the world's wildlife is of growing concern to more and more human beings. But this is a race against time. Will people and governments be aware of the before too much is lost? The future of world's life is still a big question?